NANOTECHNOLOGY SCIENCE AND TECHNOLOGY

MICROFLUIDICS

CONTROL, MANIPULATION AND BEHAVIORAL APPLICATIONS

NANOTECHNOLOGY SCIENCE AND TECHNOLOGY

Additional books in this series can be found on Nova's website under the Series tab.

Additional e-books in this series can be found on Nova's website under the e-book tab.

RECENT TRENDS IN BIOTECHNOLOGY

Additional books in this series can be found on Nova's website under the Series tab.

Additional e-books in this series can be found on Nova's website under the e-book tab.

NANOTECHNOLOGY SCIENCE AND TECHNOLOGY

MICROFLUIDICS

CONTROL, MANIPULATION AND BEHAVIORAL APPLICATIONS

STEFANO PANZARELLA
AND
WALTER MARONI
EDITORS

nova publishers
New York

Library of Congress Cataloging-in-Publication Data

ISBN: 978-1-62417-345-5

Published by Nova Science Publishers, Inc. † New York

CONTENTS

PREFACE

In this book, the authors discuss the control, manipulation and behavioral applications of microfluidics. Topics include dielectrophoresis based droplet microfluidic devices for on-chip bioassays; microfluidic biosensor systems for cell biology and drug discovery; controllable microfluidic generation of monodisperse multiple emulsion droplets; an alternative method to predict wettability for microfluidics; integration of capacitive micromachined ultrasound transducers to microfluidic devices; selection of easily accessible PCR-and bio-compatible materials for microfluidic chips; and electrokinetic manipulation of biomolecules.

Chapter 1 – Dielectrophoresis (DEP) is one of the principle electrokinetic methods of handling and processing micro-sized dielectric particles and biological cells immersed in aqueous media.

Example applications include the utility of DEP to affect separation of cells based on polarization, cell trapping/concentration, cell aggregation, etc. More recently the DEP phenomena/effect has been exploited at microscopic scales and successfully demonstrated to be capable of electro-actuating aqueous polarizable liquids on top of surfaces. This liquid DEP actuation capability has been utilized for novel dispensing, handling/manipulation of ultra-low volume aqueous suspension droplets in a rapid and automated fashion. This DEP assisted droplet dispensing capability and its integration with subsequent downstream manipulation capability suggests numerous means of applying it for lab-on-a-chip based biological/bio-chemical assays, employing ultralow sample/reagents.

The aim of this book chapter is to provide a concise review of the DEP phenomena, focusing primarily on its application to the actuation of polarizable aqueous media. In the first chapter section, the authors examine from a theoretical standpoint, the fundamental requirements of DEP assisted liquid actuation and the underlying concepts behind its various applications for lab-on-a-chip (LOC). The authors furthermore discuss how the dispensed daughter droplets, using another set of coplanar electrodes and AC excitation frequencies, can be individually transported to specific locations on the surface. Since the proposed utility of the devices focuses on bio-chemical samples and reagents, the authors specifically examine the impact of sample electrical and fluidic properties on the L-DEP actuation, droplet dispensing and transport.

This will be followed by a discussion of various approaches available for the fabrication of DEP surface microfluidic devices for reliable and repeatable operation.

Having laid a foundation for DEP based microfluidic devices, the article focuses on demonstrating various capabilities of the ultrafine droplet dispensing and manipulation scheme. The versatility of the scheme is illustrated by dispensing homogeneous and multilayered droplets (emulsions/vesicles) as well as controlled dispensing of functionalized micro-beads and nanoparticles. Some frequently used bioassays, such as DNA quantification, selective DNA hybridization, synthesis of bilayer vesicle, membrane based bio-detection assay and 'bead based' DNA/RNA hybridization detection assay, are demonstrated leveraging DEP surface microfluidic technology to facilitate miniaturized chip based bio-detection/analysis. Other novel technology capabilities include the precision dispensing of monolayer/bilayer vesicles which can serve as scaffolds for the assembly of artificial cells. The concluding section examines the proposed improvements and future development and application of the DEP droplet technology.

Chapter 2 – The ability to translate drug-target interactions into cell phenotypic signatures has made label-free biosensors become a powerful platform for the molecular delineation of receptor signaling and drug pharmacology in native cells. Microfluidics enables miniaturization of assays, increases experimental throughput, and allows precise control of cell phenotypes, molecular concentrations and drug-target interactions. Integration of microfluidics with label-free biosensors would offer additional dimensions for investigating the cellular and molecular mechanisms of receptor signaling and drug actions down to single cell level. This chapter first reviews key aspects of microfluidics for studying cell biology, and discusses the recent advances and applications in cell biology and drug discovery of microfluidic biosensor devices.

Chapter 3 – Multiple emulsions, or "emulsions of emulsions", are complex nested liquid systems, in which dispersed droplets contain smaller droplets inside. They are widely used to encapsulate active ingredients in myriad applications, including drug delivery systems, foods, cosmetics, chemical separations, and syntheses of microspheres and microcapsules. Accurate control of size monodispersity and internal structure are critical for the versatility of such emulsions, because these attributes allow precise manipulation of the encapsulation of substances and the structure of synthesized materials. Compared with other emulsification techniques, such as multistep bulk emulsification and membrane emulsification, recently developed microfluidic technique shows much greater power in controlled generation of monodisperse multiple emulsion droplets. In this Chapter, droplet microfluidic device with highly scalability for generating monodisperse multiple emulsion droplets are firstly introduced, from which the readers can see the precise control on both the size and number of the inner droplets in multiple emulsions; then, based on this device, novel devices for controllable encapsulating droplets with different components in each level of multiple emulsions are introduced; the size, number, and ratio of the multicomponent inner droplets are under excellent control, which largely extend the emulsion structures and benefit the application of multiple emulsions.

Finally, the application of these multiple emulsions are demonstrated by template synthesis of functional microcapsules with highly monodispersity, controlled inner structures and encapsulation characteristics.

Chapter 4 – Wettability properties inside the microchannels of microfluidic devices have received increasing attention over the past few years due to their important roles in controlling liquid flow. Conventionally, the wetting properties of a surface are determined by contact angle measurement where a liquid drop is placed on the surface and the angle between the solid surface and liquid-vapour interface is measured. The size of the liquid drop on the surface is often larger than the size of components or features in microfluidic systems.

This renders conventional methods for measuring contact angle incapable of delivering the information required. This chapter presents a brief review of the authors' recent research in the development of an indirect approach for predicting contact angle by the use of surface analytical methods such as time-of flight secondary ion mass spectrometry (ToF-SIMS) and x-ray photoelectron spectroscopy (XPS). Two examples have been shown here. One is to predict micron-scale surface wettability maps of substrates by exploitation of surface chemical imaging using SIMS. Using the surface sensitive and chemically specific spectroscopy available with SIMS, it was possible to obtain information on species that are related to wetting behavior and their distribution over the surface. Another example has demonstrated that XPS, similar to SIMS, has the potential to give precise surface chemical information and provide insights into the quantitative assessment of the surface functional groups, allowing the prediction of surface wettability. The alternative method described here is of direct benefit to industries where wettability is crucial but contact angle measurements cannot be directly acquired on more complicated or flexible surfaces.

Chapter 5 – The design and manufacturing flexibility of capacitive micromachined ultrasound transducers (CMUT) makes them attractive option for integration with microfluidic devices both for sensing and fluid manipulation. CMUT concept is introduced here by presenting simplified equivalent circuit model; while finite element analysis (FEA) and near-field CMUT operation experiments were used as the main methods of research. The transient time FEA was used to simulate the transmission and reception of the short pulse and then analysis of the CMUT membrane motion in the time and frequency domains was made. 200 nm finite element spacing and 0.2 ns sampling time was used during the simulations for increased precision and reliability of the results. Surface micromachined CMUT devices with two-phase interdigital arrangement of the transducer elements were fabricated for experiments. The microchannels of different heights were modeled by attaching the adhesive tape of various thicknesses at the sides of the CMUT devices to form the bead for the liquid. The drop of the transformer oil was put in the microchannel before attaching the upper microchannel wall. Afterwards the electromechanical impedance of CMUT was measured with the network analyzer to evaluate the CMUT operating regime.

The finite element analysis showed that when the microchannel cross-section dimensions are comparable with the CMUT cell dimensions, the transducer is subjected to the pressure waves bouncing between the microchannel walls and comparatively compliant CMUT membranes. This situation causes the distortion of the CMUT operating regime and has to be accounted during the design and use of CMUT-coupled microfluidic devices. Corrections for the amount of the near-field distortions are to be made already during the design phase.

The authors found significant relationship between the CMUT membrane motion characteristics and the microchannel height. For example: CMUT having 5 MHz central frequency in immersion exhibits the broad band frequency spectra when in normal operating conditions, but the central frequency decreases by 60-70 % with decreasing the microchannel height from 400 microns to 2 microns, and the frequency band becomes narrower. It is also shown here how the simulated results can be experimentally verified.

Chapter 6 – Conventional fabrication of microfluidic chip is a complicated and time, effort and material consuming process. Consequently, due to high expenses, it has poor applicability for performing mass biological analysis by microfluidics. In this study, the authors report several measures to make simple, low-cost and rapidly fabricated two-dimensional (2D) and three-dimensional (3D) microfluidic chips for biological analysis. The authors employed various easily accessible and polymerase chain reaction (PCR) compatibility verified materials in the process of design, patterning and bonding. Firstly, they established a simple method for evaluating the PCR compatibility of various common materials, as DNA and proteins are common components or targets in biological analysis. These materials employed in fabricating microfluidic chips, include silicon, several kinds of silicon oxide, glasses, plastics, wax, and adhesives. Two-temperature PCR was performed with these materials to determine their PCR-inhibitory effect. In most cases, the addition of bovine serum albumin (BSA) effectively improved the reaction yield. The authors also studied the individual PCR components from the standpoint of adsorption. Most of the materials did not inhibit the DNA, although they noticeably interacted with the polymerase. The authors' results provide an overview of materials that are PCR-friendly for fabricating microfluidic devices. The PCR reaction, without any additives, performed best with pyrex glass, and it performed worst with PMMA. Then, based on the results of the PCR-compatible screening test, the authors concentrated on fabricating microfluidic chips using easily accessible materials, such as paper and wax. The authors successfully shortened the whole fabrication process for complicated 2D and 3D microfluidic chips from the usual several days into several minutes by using easily accessible materials and methods. They demonstrated that these types of chips have good bio-compatibility enabling their applications in many biological analyses such as PCR and DNA capillary electrophoresis. The authors also dramatically simplified the process of fabricating complicated 3D chips by using cyanoacrylate-based resin or adhesive wax as the bonding material to integrate chip materials such as paper, glass slides or other polymer films. With this process, a 3D microfluidic chip is achievable by vacuating and venting the chip. And the bio-compatibility and applicability of the paper-based 3D microfluidic chip was verified in the applications such as PCR, *HeLa* cell electroporation and the chemotaxis of *E. coli*. In all, the authors' PCR-compatibility test provides good hints as to the selection of materials for microfluidic chips employed in biological analysis. Base on this, the authors succeeded in fabricating complicated microfluidic chips using easily accessible materials that may make it much more applicable than conventional methods.

Chapter 7 – Electrokinetics in microfluidic system provides effective molecular manipulation technique in micro/nano domain, which matches the length scale of biomolecules. A lot of demonstrations have been reported and shown the manipulation of different biomolecules using different kinds of electrokinetic phenomena. This chapter reviewed the most commonly used electrokinetic phenomena including DC electrophoresis, AC electrophoresis, DC electroosmosis, AC electroosmosis, electrowetting, and opto-electrically induced electrokinetics. The principle behind and recent developments on the electrokinetic manipulation will be discussed. An updated and systematic in-depth discussion is provided in the field of electrokinetic manipulation.

In: Microfluidics
Editors: Stefano Panzarella and Walter Maroni
ISBN: 978-1-62417-345-5

Chapter 1

DIELECTROPHORESIS BASED DROPLET MICROFLUIDIC DEVICES FOR ON-CHIP BIOASSAYS

Karan V. I. S. Kaler and Ravi Prakash
Biosystems Research and Applications Group (BRAG),
University of Calgary, Calgary, Canada

ABSTRACT

Dielectrophoresis (DEP) is one of the principle electrokinetic methods of handling and processing micro-sized dielectric particles and biological cells immersed in aqueous media.

Example applications include the utility of DEP to affect separation of cells based on polarization, cell trapping/concentration, cell aggregation, etc. More recently the DEP phenomena/effect has been exploited at microscopic scales and successfully demonstrated to be capable of electro-actuating aqueous polarizable liquids on top of surfaces. This liquid DEP actuation capability has been utilized for novel dispensing, handling/manipulation of ultra-low volume aqueous suspension droplets in a rapid and automated fashion. This DEP assisted droplet dispensing capability and its integration with subsequent downstream manipulation capability suggests numerous means of applying it for lab-on-a-chip based biological/bio-chemical assays, employing ultralow sample/reagents.

The aim of this book chapter is to provide a concise review of the DEP phenomena, focusing primarily on its application to the actuation of polarizable aqueous media. In the first chapter section, we examine from a theoretical standpoint, the fundamental requirements of DEP assisted liquid actuation and the underlying concepts behind its various applications for lab-on-a-chip (LOC). We furthermore discuss how the dispensed daughter droplets, using another set of coplanar electrodes and AC excitation frequencies, can be individually transported to specific locations on the surface. Since the proposed utility of the devices focuses on bio-chemical samples and reagents, we specifically examine the impact of sample electrical and fluidic properties on the L-DEP actuation, droplet dispensing and transport.

This will be followed by a discussion of various approaches available for the fabrication of DEP surface microfluidic devices for reliable and repeatable operation.

Having laid a foundation for DEP based microfluidic devices, the article focuses on demonstrating various capabilities of the ultrafine droplet dispensing and manipulation scheme. The versatility of the scheme is illustrated by dispensing homogeneous and multilayered droplets (emulsions/vesicles) as well as controlled dispensing of functionalized micro-beads and nanoparticles. Some frequently used bioassays, such as DNA quantification, selective DNA hybridization, synthesis of bilayer vesicle, membrane based bio-detection assay and 'bead based' DNA/RNA hybridization detection assay, are demonstrated leveraging DEP surface microfluidic technology to facilitate miniaturized chip based bio-detection/analysis. Other novel technology capabilities include the precision dispensing of monolayer/bilayer vesicles which can serve as scaffolds for the assembly of artificial cells. The concluding section examines the proposed improvements and future development and application of the DEP droplet technology.

1. INTRODUCTION

Microfluidic systems handle sample volumes in range of microliter and lower while utilizing structures and geometries in micron/sub-micro region. A microfluidic system contains a combination of generic components to achieve the most essential of laboratory functionalities on the miniaturized scale. These components include:

- Method for introducing samples and reagents, i.e. Dispensing and sample loading
- Method for moving samples to targeted regions, i.e. transport/manipulation
- Method for combining or mixing the samples/reagent
- Method for detection, sample purification/pre-screening etc…

Much of the work to date has focused on technology that can demonstrate some or all of the aforementioned capabilities. A microfluidic device that achieves all the above mentioned functionalities is essentially a miniaturized replacement to the conventional Laboratory and is often deemed as "Lab on a Chip (LOC)" device [1, 2]. Based on the methodology implemented to achieve one or more of the targeted applications, such devices have been classified as follows:

1.1. Close Channel Microfluidic Device or, Analog Microfluidic Device

These are continuous flow devices that leverage micro-channels, high pressure micro-valves, pumps and tubing arrangement to facilitate LOC functionalities [3]. The micro-channels and micro-valves are usually fabricated in PDMS using soft-lithography and assembled with the off-chip pumping/tubing arrangement to feed the liquid sample at the required high pressure [4, 5]. Since flow in micro-channels is mostly laminar given that viscosity dominates over inertia at these scales, it results in a non-turbulent mixing scheme which occurs as a result of diffusion of molecules across the interface between the fluids. So, the mixing achieved in micro-channels requires long channel lengths.

Also, the requirement of bulky off-chip components hinders the possibility of multiplexing and microfluidic large scale integration (MF-LSI). Large scale integration (LSI) refers to the ability to integrate numerous functional components on one device to achieve complex tasks. In closed channel microfluidics, MF-LSI refers to microfluidic chips with numerous micro-mechanical valves that can transport and manipulate numeral fluidic samples in parallel and automated fashion [6]. MF-LSI and high throughput screening (HTS) [7, 8] are two more recent advancements in close-channel microfluidics. Different liquid handling techniques have been attempted to overcome certain drawbacks of the conventional close channel method. Quake and co-workers have demonstrated design and fabrication protocols for multiplexed arrays of monolithic valves which are addressed in a controlled, multiplexed fashion, enabling MF-LSI and HTS [8, 9]. This is achieved by using multilayer soft-lithography technique in two overlapping layers. The 'flow' layer contains the microfluidic channels and the 'control' layer harbors the channels required to actuate the valves and control the flow. In another recent approach, centrifuge-based close channel microfluidic platforms, also known as "Lab on a Disc" are devised where the underlying physical principles of centrifugal pumping is utilized in context of microfluidic systems and the various centrifuge fluidic functions, such as valving, mixing, sample splitting and separation, are introduced [10]. But, even with the advent of multiplexing in close-channel microfluidic, sensor integration and bulky off-chip component remain as major unaddressed drawbacks of this technology.

1.2. Surface Microfluidic Device or, Digital Microfluidic Device

In the case of surface microfluidic (SMF) devices, precision sample droplets, controlled over patterned surfaces are used to dispense and manipulate various samples and reagents. Several droplet actuation and manipulation schemes have been utilized in recent years. Some of these schemes such as: a) droplet manipulation by controlling the surface wettability (surfaces coated with self-assembled monolayer) [11]; b) use of surface temperature gradients for sample droplet manipulations [12] are deemed as passive since there is no active switching of external fields. Passive droplet transport schemes are slow and restrictive and hence they have mostly been replaced by more common active digital microfluidic technologies where the electric field is actively controlled to achieve droplet manipulations. One of these schemes is based on Surface acoustic waves (SAW), where patterned, inter-digitated (IDT) electrodes are used to create controlled surface acoustic vibrations which control the transport of individual sample droplets [13]. Additional schemes have also been proposed where droplets are manipulated using piezo-electric effects and by using acoustic waves [14], but the two more popular SMF actuation approaches are based on use of patterned, programmable electrodes which are suitably energized using an external AC/DC supply to overcome the fluidic resistance to achieve droplet manipulation. The two popular schemes are: Electrowetting (EW) or, Electrowetting-on-Dielectric (EWOD) and Dielectrophoresis (DEP).

Electrocapillarity, first demonstrated by Gabriel Lippmann in 1875 [15], is quoted as the basis for the modern day electrowetting devices (EW). In such EW devices, low voltage (0-100 Vpp), low frequency (DC-1 kHz) electric field are implemented to perturb the interfacial equilibrium of the liquid-surface or liquid-liquid interface and hence control the shape of a sample/reagent droplets. Suitably designed and fabricated electrode arrays can be utilized to facilitate dispensing, splitting, transport, merging and multiplexed screening of sample/ reagent droplets and henceforth achieving most of the requirements of a LOC devices [16, 17,

18]. Although, EW or, EWOD actuation schemes provide more flexible droplet handling capabilities, its speed and proficiency is controlled by their fluidic properties.

Dielectrophoresis (DEP) was originally defined by H. A. Pohl in 1958 [19] as the underlying concept for manipulating dielectric media under the influence of spatially non-uniform electric field. However, the first experimental demonstration of DEP goes further back to Pellat's demonstration of wall-less flow systems in 1894 [20]. Melcher's dielectric siphon in 1971 [21] was another experiment implementing the non-uniform electric field to manipulate fluidic dielectric samples. Both these observations, which are summarized in the theory section, leveraged DEP on macroscale and required extremely high actuation voltages (~ 30 kVrms) but when the dimensions are scaled down to micron or sub-micron scale as in case of microfluidics, the required actuation voltages can be substantially lowered due to a favorable scaling relation (see Theory section) [22]. DEP has now emerged as a successful tool to manipulate samples for on-chip, microfluidic applications.

The phenomenon of DEP has since been implemented in flow separation [23], classification and segregation of bio-particles in various fluidic media (particle-DEP) [24, 25, 26] and for dispensing and subsequent manipulation of ultrafine (in nL – pL volume range), sample/reagent droplets [27, 28, 29], commonly known as Liquid-DEP (L-DEP). This chapter focuses on the various characteristic attributes of L-DEP and covers its advancement as a suitable microfluidic tool for chip-based bio-applications.

Electrostatic droplet actuation or, D-DEP: The electrostatic droplet actuation scheme, also termed as droplet-DEP (D-DEP) [29, 30, 31] has been used to achieve subsequent manipulation of L-DEP actuated droplets in the reported work. Electrostatic droplet actuation scheme was first demonstrated by Washizu and co-workers [30] in 2004.

The actuation scheme uses fishbone shaped electrodes, energized at low frequency AC voltage/or voltage pulse in order to manipulate droplets on open surfaces. The underlying concept of this actuation methodology is explained in a later section.

2. THEORY

2.1. Liquid Dielectrophoresis- The Phenomena and Its Characteristic Features

L-DEP- Theoretical background

As described earlier, DEP is an electrokinetic phenomenon that dictates the influence of an external, spatially non-uniform electric field, on a dielectric medium. In case of fluidic samples, the DEP force acts on the polar fluidic molecules and tends to redistribute the fluidic mass towards the regions of high field intensity. This electrokinetic re-distribution is in practice opposed by the fluidic surface tension (F_γ) and the viscous damping force (F_μ) and was termed as liquid-dielectrophoresis (L-DEP) by Jones and co-workers [27, 32]. Essentially, in case of L-DEP actuation, polarizable liquid is manipulated by this ever changing hydrostatic equilibrium of the DEP force and the fluidic resistance and is well contained and furthermore shaped by the spatially non-uniform electric field.Even before the phenomenon of L-DEP was officially termed and defined, there were intuitive demonstrations where electric field based ponderomotive force was used to demonstrate electrocapillarity effect. This includes Pellat's classical demonstration in 1894 [20] which provides a basis for wall-less flow of dielectric liquids. In his experiment (shown in Figure 1(a)), Pellat used two plane parallel electrodes, oriented vertically at a separation s and partially immersed in a dielectric liquid (density ρ and permittivity ε). When a voltage V is applied across the electrodes, in order to attain a new hydrostatic equilibrium, liquid between the two electrodes rises to a height h given by,

$$h \approx \frac{(\varepsilon - \varepsilon_0)V^2}{2s^2 \rho g} \tag{1}$$

where, g = 9.81 m/s^2 is the acceleration due to gravity and ε_0 is the permittivity of space.

Another example of wall-less flow structure is the dielectric siphon [21] which was demonstrated by Melcher and co-workers in 1971 and is shown in Figure 1(b). In this set-up, they had two closely spaced (s < 1 cm) parallel electrodes running between two reservoirs placed at different levels and containing dielectric liquids.

At sufficiently high voltage (30 kVrms at 400 Hz), a fluidic connection was successfully established between the two reservoirs. Since both the demonstrations required impractically high voltages in order to achieve fluidic manipulation at macroscale, the use of DEP was less favored as compared to conventional hydraulic and pneumatic components (valves and pressure pumps). However, in more recent attempts by Jones and co-workers (1975), where they used DEP to maneuver dielectric fluids at miniaturized scales [21, 22], it was successfully demonstrated that excessive voltage requirements and issue of electrical breakdown can be avoided at microscale applications mostly because of the favorable scaling laws.

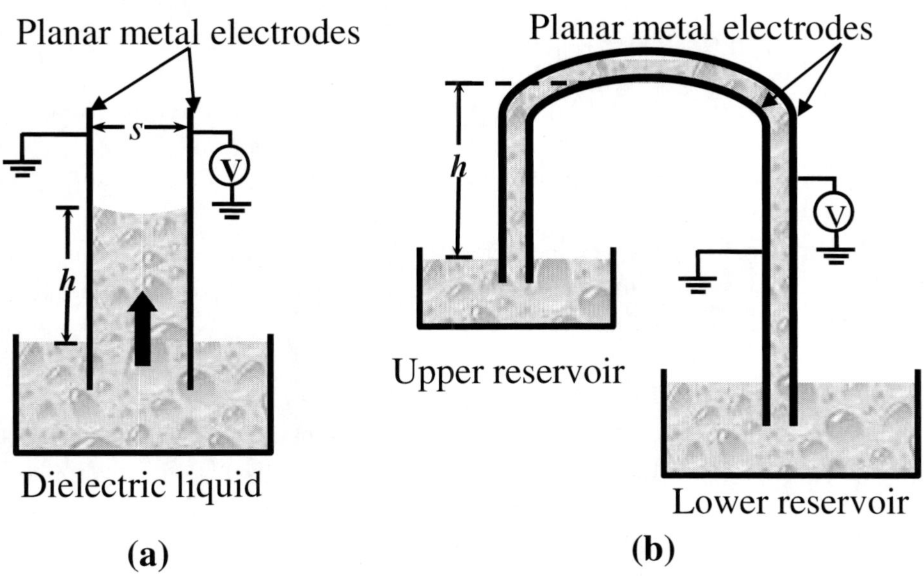

Figure 1. Schematics showing (a) Pellat's classic demonstration of a wall-less flow structure; (b) Dielectric siphon developed by Melcher and co-workers.

Table 1. Rheological and electrical properties of used liquid samples

Sample name	Composition (% by Vol. of glycerol)	Jet Capacitance (C_{eff} in pF)*	Viscosity (cSt)*	Surface Tension (dyne/cm)*	Conductivity (μS/cm)
1	0	553.60	1	73	0.65
2	12.5	539.76	2	71	2.4
3	25	519	4	68	6.0
4	50	477.48	8	65	15.5

* Jet capacitance, viscosity, surface tension and conductivity values referred from [35, 38].

The reason behind this can be better understood once the concepts of L-DEP are elaborated in the next section. In 2001, Jones and co-workers [32] developed an L-DEP based droplet dispensing scheme by using miniaturized co-planar metal electrodes which were micro-fabricated on silicon/glass substrates.

Once successfully implemented in fluidic manipulation at microscale, L-DEP based scheme have since been modified and state-of-art microfabrication techniques have been used to devise DEP based SMF LOC devices [29, 33, 34].

The two most common scenarios of DEP based dielectric manipulation are often classified as: particle-DEP, where a dielectric micro-particle is moved, screened or, sorted by means DEP forces and Liquid-DEP (L-DEP) where polarizable/dielectric liquids are dispensed and further manipulated using DEP forces. Due to its favourable scalability in the micro-dimension [22], DEP has developed into a suitable means of sample handling and processing in microfluidics where its applications range from particle/cell sorting [23, 24, 25], homogenous and uniform droplet dispensing [27], single emulsion droplet dispensing [35], and controlled volume dispensing [36], uniform deposition of macro-molecules and micro-sized particles in nanoliter-picoliter sample aliquots (daughter droplets) [37, 38].

In the next segment, fundamental concepts of L-DEP, Jones's simplified lumped capacitance model and its modification to accommodate more complex liquid actuations are explained. Experimental results obtained from the various L-DEP based liquid actuation methods are utilized to validate the theoretical models. Aqueous samples used during these L-DEP actuations are reported in Table 1.

Lumped capacitance model for homogenous L-DEP actuation

A simple L-DEP micro actuation scheme consists of a pair of coplanar metal electrodes of widthw and gap g, insulated on top with a thin dielectric layer with a hydrophobic top surface coating, to prevent sample electrolysis and facilitate droplet dispensing and their subsequent manipulation (fabrication details in next section). When a parent sample droplet is placed on one end of the electrode pair and the electrodes are electrically energized by a sufficiently high AC voltage (typically ~ 200-500 Vpp at 50 kHz-1 MHz), a liquid jet is ejected from the parent droplet and is rapidly conveyed along the electrode pair, covering the entire electrode length within a few milliseconds (Figure 2).

In order to prevent evaporative sample loss, all liquid actuations are carried out in a low viscosity (5-10 cSt) silicone oil bath. The profile of the actuated liquid jet is actively controlled by the shape of the electrode arrangement, top surface coating and fluidic properties of the liquid sample. The dynamics of such L-DEP actuation of homogenous liquid jet has been analyzed based on a lumped parameter model [32].

Figure 2 shows the physical model for a liquid jet protruding from a parent sample droplet during uniform L-DEP actuation. In this model, the DEP force (F_{DEP}) is opposed by surface tension (F_γ) and viscous force (F_μ) and under the assumption of a uniform hemi-cylindrical jet, the force-momentum equation is described by:

$$\frac{dP}{dt} = F_{DEP} - F_\gamma - F_\mu \quad (2)$$

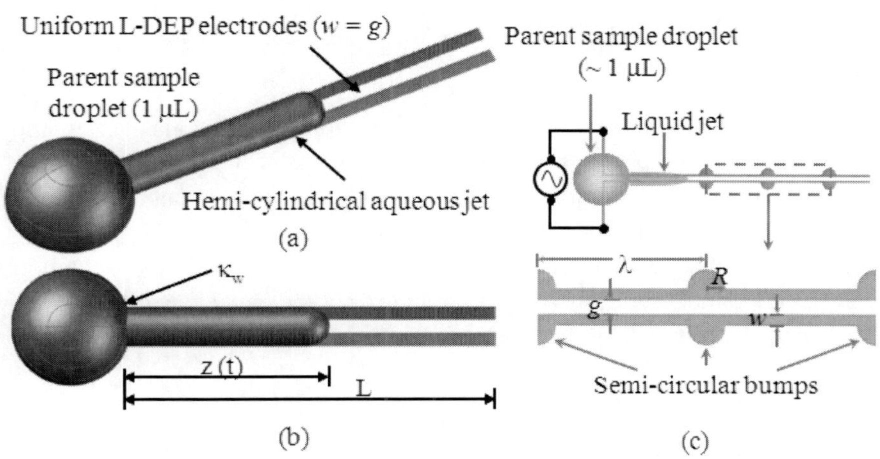

Figure 2. (a) 3-d model of a uniform L-DEP actuation; (b) Top view of the model; (c) Schematic details of a simple uniform L-DEP electrode scheme.

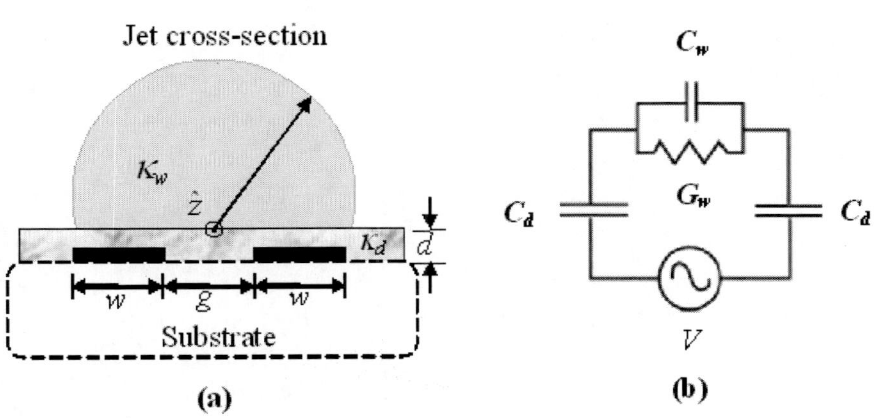

Figure 3. (a) X-sectional view of an aqueous jet during L-DEP actuation; (b) RC equivalent circuit for the capacitive coupled electrode/actuation scheme on the SMF chip.

The rate of change of momentum term incorporates the time-dependent finger length (z(t)) and assuming that the parent droplet remains unperturbed during the actuation, the governing differential equation can be expressed as:

$$\frac{\pi\rho R^2}{2}\frac{d}{dt}\left[z\frac{dz}{dt}\right]=F_{DEP}-F_{\gamma}-F_{\mu} \tag{3}$$

where, z is the actuated jet length. The assumption that the liquid jet profile is uniform hemi-cylindrical, allowed the use of a lumped parameter model, for calculating F_{DEP}.

In this approach, lumped total system capacitance (*C(z)*) is used to evaluate the DEP force using principle of virtual work [56]. The lumped total system capacitance term can be expressed as;

$$C(z)=C_{drop}+C_{c}+C_{uc} \tag{4}$$

where C_{drop} is the capacitance term contributed from the electrode pair overlapping with the parent droplet, C_c is the lumped capacitance term contributed from the electrode area covered by the moving jet at instant t and jet length z and C_{uc} is the lumped capacitance term contributed from the uncovered electrode area.

Figure 3 shows an equivalent circuit model representing the liquid actuation scheme. Based on such equivalent circuit models, one can extract the capacitance terms and write the total lumped capacitance as;

$$C(z)=\frac{z}{1/\kappa_w c_{air}+2/c_d}+\frac{L-z}{1/c_{air}+2/c_d}+C_{drop} \tag{5}$$

L is the total length of the electrodes (not covered by the parent droplet), κ_w is the relative dielectric constant of the aqueous sample. The term c_d is the capacitance per unit length term contributed from the dielectric coatings on top of the patterned electrodes (see Figure 3(a)) and c_{air} is the capacitance per unit length of the hypothetical hemi-cylindrical shape covering the electrode area. The two capacitance terms are as follows;

$$c_d=\frac{K_d\varepsilon_0 w}{d};c_{air}=\varepsilon_0\frac{K(1-k^2)}{2K(k^2)} \text{ and } k=\frac{g}{2R} \tag{6}$$

$K(k^2)$ and $K(1\text{-}k^2)$ are complete elliptical integrals of the first kind and are used to formulate the lumped capacitance term contributed from the hemi-cylindrical shaped liquid jet profile. d is the thickness of the top dielectric layer while R is cross-sectional radius of the liquid jet profile and can be evaluated as: $R = w + g/2$. This lumped analysis remains accurate for $d << w, g$. Employing the principle of work for the lumped system, the DEP force can be calculated as;

$$F_{DEP} = \frac{\partial W_e}{\partial z}\bigg|_{V=const} = \frac{V^2}{2}\frac{dC(z)}{dz}; W_e = C(z)\frac{V^2}{2} \tag{7}$$

Using the total system capacitance given by Eqn. 5 in Eqn. 7 gives the lumped expression for F_{DEP} as;

$$F_{DEP} = \frac{(K_w - 1)c_d^2 c_{air} V^2}{2\left(c_d + 2K_w c_{air}\right)(c_d + 2c_{air})} \tag{8}$$

The expression for the DEP force term remains constant for a uniform L-DEP actuation. The surface tension force (F_γ) acts to impede the increase in the fluidic surface area during jet actuation and under the assumption of constant and uniform jet cross-section, remains constant and given as a product of semi-circular perimeter;

$$F_\gamma = \pi R \gamma \tag{9}$$

The fluidic viscous force which resists motion of the liquid jet can be expressed as the product of the jet projection area at an instant t, given as $2Rz(t)$ and the shear stress τ_μ which, for the Newtonian fluids, can be expressed as;

$$\tau_\mu = \mu\frac{\partial v}{\partial y}; F_\mu = 2Rz\tau_\mu \tag{10}$$

μ is the dynamic viscosity of the fluidic sample and $\frac{\partial v}{\partial y}$ is the velocity gradient (with no slip boundary condition at the surface) in the jet profile.

Reynolds Number$\left(\mathrm{Re}\right) = \frac{\rho v L}{\mu}$, gives the ratio of inertial force to viscous force for a particular flow scheme and is used to investigate whether the flow is laminar or turbulent (Re < 2300 corresponds to a laminar flow). It is important to note that the characteristic Reynolds number for flow in micro-channels (open or closed) are fairly low (10-100) due to the micron scale flow sections, as a result of which flow in micro-channels and also during L-DEP actuation are always laminar with a velocity profile governed by Poiseuille expression for laminar flow [32].

The knowledge of the jet velocity profile allowed rewriting of the velocity gradient in Eqn. 10 in terms of a characteristic constant (D) also referred in literature as viscosity coefficient and the averaged instantaneous liquid jet velocity, $\left(\frac{dz}{dt}\right)$. The lumped expression for shear stress and viscous force is reported as:

$$\tau_\mu = \mu\frac{\partial v}{\partial x} = \frac{D\mu}{R}\frac{dz}{dt}; F_\mu = (2\mu D)z\frac{dz}{dt} \tag{11}$$

The coefficient D depends on velocity profile of the liquid jet and thus essentially on the electrode dimensions.Substituting viscous force expression from Eqn. 11 in Eqn. 3 and rearranging gives the governing differential equation (DE) as;

$$\frac{d^2z^2}{dt^2} + \frac{1}{T_\mu}\frac{dz^2}{dt} = \frac{F_{DEP} - F_\gamma}{\frac{\rho\pi R^2}{4}} \tag{12}$$

$T_\mu = \dfrac{\rho\pi R^2}{4\mu D}$ is a characteristic time constant that delineates the boundary between viscous and inertia dominated behavior. The DE is solved with the initial condition of: z(t = 0) = 0 and the actuation voltage *V* being applied at t = 0 s and reported as;

$$z(t) = \sqrt{\frac{T_\mu(F_{DEP} - F_\gamma)}{\rho\pi R^2/4}[t + T_\mu(e^{-t/T_\mu} - 1)]} \tag{13}$$

The solution can be simplified for the viscosity and inertia dominant domains as:

$$z(t) \approx \begin{cases} \left(\dfrac{(F_{DEP} - F_\gamma)}{\rho\pi R^2/2}\right)t, & t \ll T_\mu \\ \dfrac{(F_{DEP} - F_\gamma)}{\rho\pi R^2/4}t, & t \gg T_\mu \end{cases} \tag{14}$$

For $t \ll T_\mu$, the dynamics is inertial controlled and is generally observed in case of large or, non-uniform electrode schemes and actuation of low viscosity oils where a linear time dependence is observed. For $t \gg T_\mu$, the jet dynamics is viscosity controlled and this condition has been observed in actuation of aqueous samples and reagent on micron sized L-DEP electrode schemes. The extracted result (Eqn. 14) is applied to L-DEP actuation of different homogeneous glycerol-DI water mixtures (Table 1) and shown in Figure 4.

Thus far, the lumped model has been used to predict the transient behavior of homogenous liquid jets. This section deals with L-DEP actuation schemes which result in formation of non-uniform, tapered liquid jets leading to formation of variable volume daughter droplet arrays. A tapered jet is produced as a result of L-DEP actuation over tapered L-DEP electrode scheme which essentially consists of L-DEP electrodes which are not parallel but converging (tapered) or diverging (tapered up) electrode geometry.

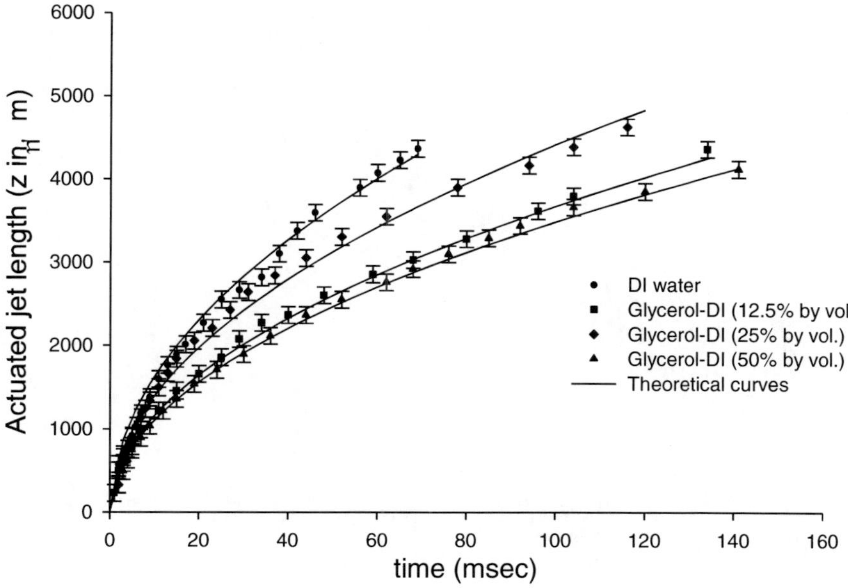

Figure 4. Experimental validation of the predicted $t^{0.5}$ uniform liquid jet actuation.

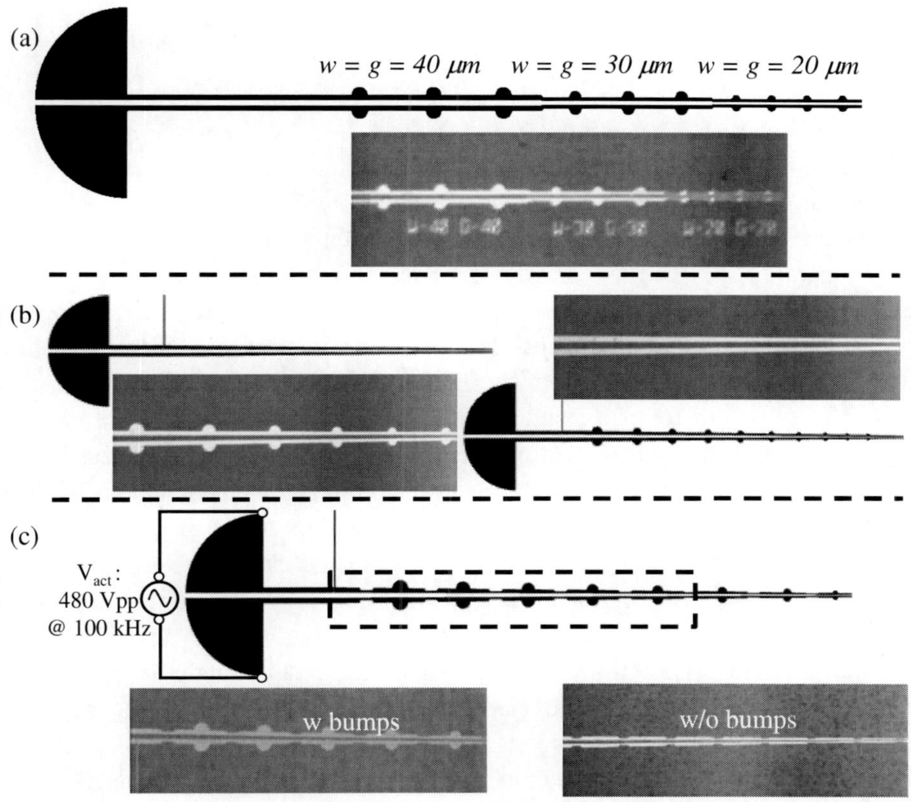

Figure 5. Screenshots and schematics showing different tapered L-DEP electrode morphologies. a) 3-step tapered L-DEP electrode scheme; b) continuous tapered L-DEP electrode scheme; (c) Pinched continuous tapered electrode scheme.

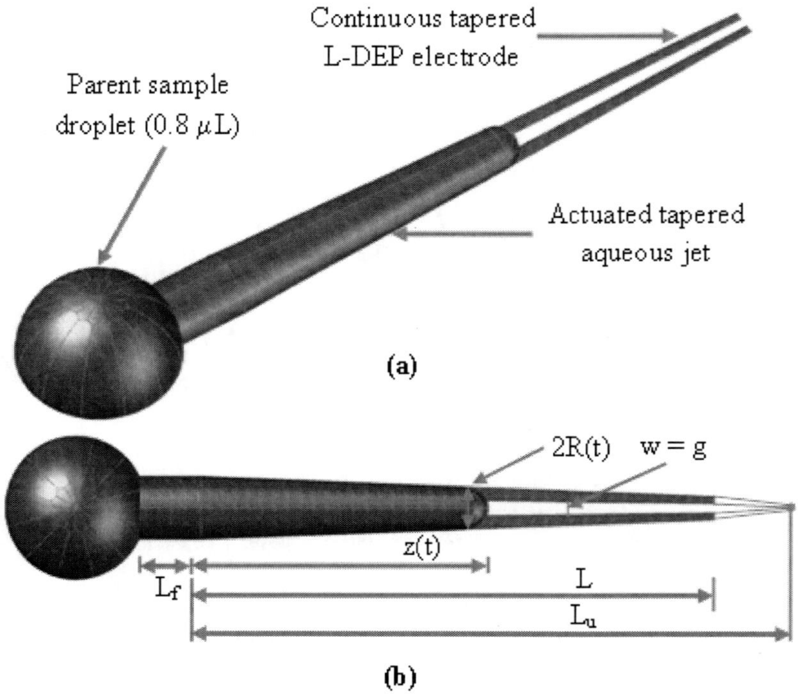

Figure 6. (a) 3-d schematic of a tapered L-DEP actuation; (b) Top view of the model.

When such an electrode scheme is used for L-DEP actuation, a non-uniform tapered jet is formed and the dynamics of such a tapered jet is expected to be more cumbersome than the uniform jet scenario.

The concept of tapered L-DEP was first introduced by Chugh and Kaler [34] when they implemented an L-DEP electrode scheme, illustrated in Figure 5(a), consisting of three segments of parallel but stepped down electrode pairs (w-g: 40-40; 30-30 and 20-20), resulting in a 3-step tapered L-DEP scheme which can dispense three sets of droplets with different droplet volumes. Since the steps of the step tapered electrode scheme are uniform L-DEP electrode pairs, it is expected that the liquid jet will exhibit $t^{0.5}$ dependence but, the profile of the plots will differ as a result of differing initial conditions for the moving jet in each of the three uniform L-DEP electrode sections. In this section, we discuss continuous tapered L-DEP electrode scheme where the electrodes converge in a continuous manner while maintaining the uniform (w = g) cross-section.Figure 6 shows 3-D model for a continuous tapered L-DEP actuation scheme. In this continuous tapered electrode design, the width and gap of the electrode pair is decreased while maintaining w = g, along the entire electrode length. In order to practically utilize such a tapered electrode design with electrode length which is significant for droplet dispensing applications, the taper angle has to be kept small (~1°). For such a continuous tapered electrode scheme, F_{DEP} and F_{γ} can no longer be represented by constant lumped expressions. Expressions for the modified force terms and the governing non-linear differential equation (NLDE) are generated and solved in order to explain the dynamics of a tapered liquid jet, actuated over a continuous tapered L-DEP electrode [36]. The expression for mass momentum, rate of change of mass momentum and the three force terms are recalculated in terms of indefinite integral expressions, for the new actuation scheme. The momentum term for the tapered actuation is expressed as:

$$P(z,t)=\left(\int_0^Z \frac{\rho\pi R(z)^2}{2}dz\right)\frac{dz}{dt}=\frac{\rho\pi R_b^{\,2}}{2}\left(\int_0^Z\left(1-\frac{z}{L_u}\right)^2 dz\right)\frac{dz}{dt} \tag{15}$$

$R(z)=R_b\left(1-\frac{z}{L_u}\right)$ is the radius of the tapered liquid jet at an instant t and jet length z whereas R_b is the jet radius at the wider end of the electrode towards the parent droplet. $L_u=\frac{W_1 L}{W_1-w_1}$ is the hypothetical length (see Figure 6) where the tapered electrodes converge to a point whereas W_1 and w_1 are the electrode widths at the wider and narrower ends respectively. During the analysis, the hypothetical full length L_u is used to normalize the actuated jet length as; $z_n=\frac{z}{L_u}$. Differentiating Eqn. 17 gives the time rate of change of mass momentum for the liquid jet as;

$$\frac{dP}{dt}=a_1(z)\frac{d^2z}{dt^2}+\frac{da_1(z)}{dz}\left(\frac{dz}{dt}\right)^2 \tag{16}$$

$a_1(z)=\frac{\rho\pi}{2}R_b^{\,2}L_u z_n\left(1-z_n+\frac{z_n^{\,2}}{3}\right)$ is a polynomial expression and a function of both z and t since z is the actuated z length at instant t during the actuation. The two lumped capacitance terms (C_c and C_{uc}) in total system capacitance expression given by Eqn. 4 can also be written as indefinite integral expressions as;

$$C_c=\int_0^z \frac{R(z)}{c1+c2R(z)}dz\,;$$

$$C_{uc}=\int_z^L \frac{R(z)}{c1+c2^{'}R(z)}dz=\int_0^L \frac{R(z)}{c1+c2^{'}R(z)}dz-\int_0^z \frac{R(z)}{c1+c2^{'}R(z)}dz \tag{17}$$

Here $c1=\frac{3d}{K_d\varepsilon_o}; c2=\frac{1}{K_w c_{air}}; c2^{'}=\frac{1}{c_{air}}$ are coefficients which depend upon the electrode geometry, thickness and dielectric properties of insulating layer (Si_3N_4) and the aqueous sample. The DEP force (F_{DEP}) is given by;

$$F_{DEP}=\frac{V^2(K_w-1)R(z)^2}{2K_w c_{air}\left(c1+c2R(z)\right)\left(c1+c2^{'}R(z)\right)} \tag{18}$$

Unlike in case of a uniform L-DEP actuation, the DEP force in a tapered jet actuation is not a constant lumped expression but rather a polynomial function of *z* or *R(z)*. The surface tension force (F_γ) now becomes;

$$F_\gamma = \pi R(z)\gamma = \pi R_b \gamma \left(1 - \frac{z}{L_u}\right) \tag{19}$$

Recalculating the viscous force (F_μ) gives;

$$F_\mu = \frac{2\mu D I_1(z)}{R(z)} \frac{dz}{dt}; I_1(z) = \int_0^z R(z)dz = z_n R_b L_u \left(1 - \frac{z_n}{2}\right) \tag{20}$$

In case of a tapered L-DEP actuation, both F_{DEP} and F_γ decreases during actuation process whereas, F_μ first increases and then decreases during actuation process.

Substituting the force and momentum terms derived above into the force-momentum equation (Eqn. 2) results into a NLDE which is essentially the governing equation for tapered L-DEP actuation.

$$z_n\left(1 - z_n + \frac{z_n^{\,2}}{3}\right)\frac{d^2 z_n}{dt^2} + \left((1 - z_n)^2\left(\frac{dz_n}{dt}\right) + \frac{z_n\left(1 - \frac{z_n}{2}\right)}{T_{\mu_b}(1 - z_n)}\right)\frac{dz_n}{dt} = F(z) \tag{21}$$

where, $T_{\mu_b} = \dfrac{\rho\pi R_b^{\,2}}{4\mu D}$, $F(z) = \dfrac{2}{\rho\pi L_u^2 R_b^{\,2}}\left(\dfrac{V^2(K_w - 1)R(z)^2}{2K_w c_{air}(c1 + c2R(z))(c1 + c2' R(z))} - \pi R(z)\gamma\right)$

represents the R.H.S of the NLDE. Analytic solutions for this NLDE were found to be less implicative than the one obtain for uniform L-DEP actuation and hence the NLDE was rearranged with the coefficients regrouped in order to produce a numerical solution. The rearranged and compact form of the NLDE can be given by:

$$z_n(1 - z_n)(z_n^{\,2} - 3z_n + 3)\frac{d^2 z_n}{dt^2} + \left(3(1 - z_n)^3\left(\frac{dz_n}{dt}\right) + \frac{3z_n\left(1 - \frac{z_n}{2}\right)}{T_{\mu_b}}\right)\frac{dz_n}{dt}$$
$$= e\left(\frac{a(1 - z_n)}{((1 - z_n) + b)((1 - z_n) + c)} - d\right)(1 - z_n)^2 \tag{22}$$

The grouped coefficients are represented using the five arbitrary constants which are

$$a = V^2\left(K_w - 1\right)R_b c_{air};\ b = \frac{3K_w c_{air} d}{K_d \varepsilon_o R_b};$$
$$c = \frac{3c_{air} d}{K_d \varepsilon_o R_b};\ d = 2\pi\gamma R_b^{\ 2};\ e = \frac{3}{\rho\pi R_b^{\ 3} L_u^{\ 2}} \quad (23)$$

The five coefficients given expressed in Eqn. 28 depend on actuation voltage, rheological and dielectric properties of fluidic sample used for actuation and structural and dielectric properties of the tapered L-DEP actuation scheme.

We now analyze the transient behavior of a tapered liquid jet using the developed numerical model.

Transient behavior of tapered jet for small z (or small actuation time): Assuming that the tapered angle is kept low, for the very initial period of actuation (t < 1 ms), we have

$$R(z) = R_b\left(1 - \frac{z}{L_u}\right) \cong R_b;as\ z \ll L_u$$

Under these circumstances, the R.H.S term of the NLDE reduces to a constant F_b given as;

$$F(b) = \left(\frac{V^2\left(K_w - 1\right)c_d^{\ 2}c_{air}}{2\left(c_d + 2K_w c_{air}\right)\left(c_d + 2c_{air}\right)} - \pi R_b \gamma\right)\frac{2}{\rho\pi R_b^{\ 2}} \quad (24)$$

Also, the L.H.S of the NLDE (Eqn. 22) reduces to,

$$z\frac{d^2 z}{dt^2} + \left(\frac{dz}{dt} + \frac{z}{T_{\mu_b}}\right)\frac{dz}{dt} \quad (25)$$

With some rearrangement, it can be easily shown that the reduced differential equation is essentially the governing equation of a uniform L-DEP actuation with R_b as the radius of the uniform liquid jet.

This analysis establishes that the initial (*t=0+*) transient behavior of a tapered L-DEP actuation will predominantly be a $t^{0.5}$ dependence, as shown in Figure 7.

The numerical solution for the NLDE (Eqn. 25) is reported in Figure 7.

The numerical model was experimentally validated by using the homogenous liquid samples reported in Table 1 and the result of the fitting exercise is shown in Figure 8.

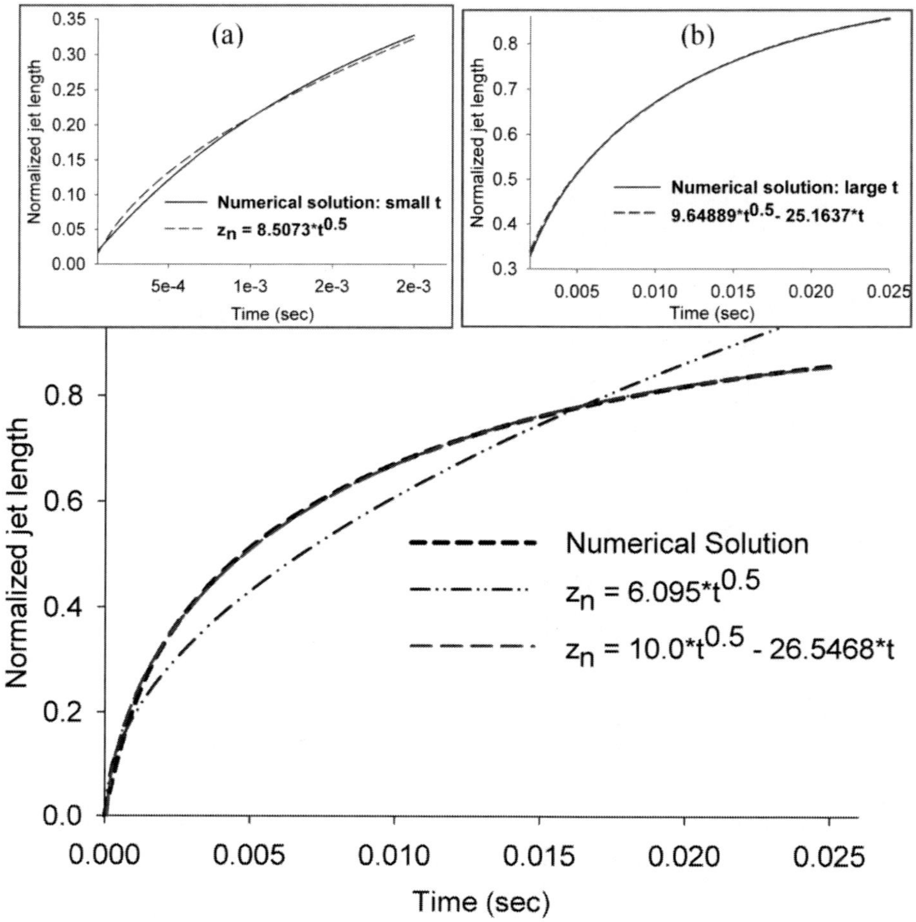

Figure 7. Solution of the numerical model for continuous tapered L-DEP actuation.

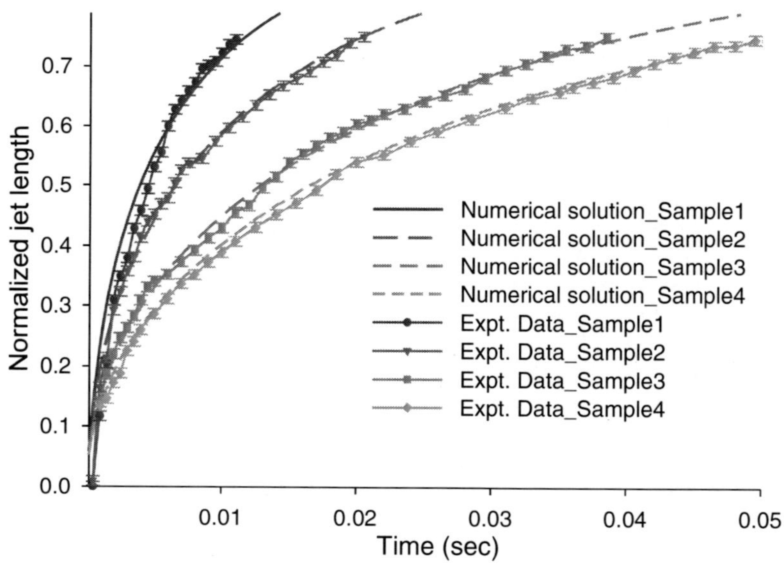

Figure 8. Experimental data fitted on the extracted numerical data from the developed model.

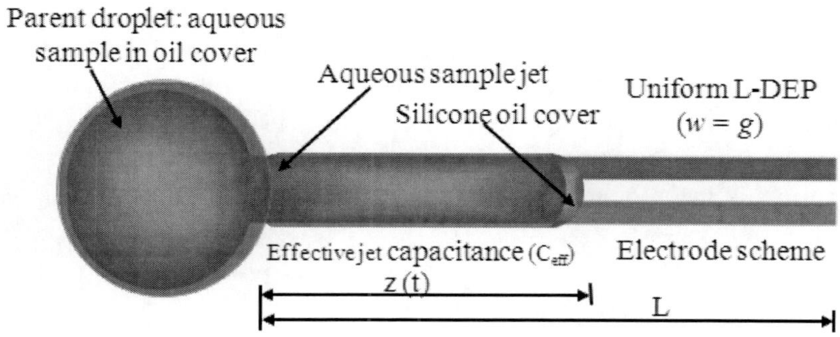

Figure 9. Model representing an L-DEP based emulsion jet actuation scheme.

Effective capacitance model for uniform, non-homogenous L-DEP actuation

In the previous sections, we discussed a lumped model governing both the uniform and non-uniform L-DEP actuation of homogeneous aqueous samples. Like other surface microfluidic actuation methods, L-DEP actuation and droplet dispensing has to be conducted under an oil bath in order to prevent volumetric loss from the dispensed sample/reagent droplets.

But the presence of an oil bath induces unwanted perturbations and contamination issues which undermine the applicability of the scheme in precision droplet dispensing for integrated and on-chip combinatorial bio-assays. The L-DEP based emulsion dispensing scheme was devised with the prime incentive to successfully substitute an oil bath by dispensing sample aliquots as emulsion droplets, individually encapsulated in comparable sized oil capsules [35].

The formation of emulsion droplets results from L-DEP actuation of an emulsion jet which is achieved by providing a comparable oil cover over the dispensed parent sample droplet as shown in Figure 9.

When the overlapping aqueous-in-oil parent droplet is subjected to sufficiently high AC voltage, an emulsion jet, consisting of an aqueous liquid jet covered in a thin oil cover is ejected from the parent droplet.

Break-up of this emulsion jet results in formation of single emulsion (SE) droplets which are essentially aqueous sample droplets individually encapsulated in comparable oil droplets and precisely positioned at specific on-chip dispensing sites.

Dynamics and other important characteristics of the emulsion jet actuation are presented in this section [35].To investigate the dynamics of an emulsion jet actuation, the approach used is essentially based upon the previously explained lumped parameter model.

In case of liquid actuation under an oil bath, the liquid jet displaces the surrounding oil media during actuation whereas for an emulsion jet, an oil cover is constantly dragged along with the aqueous jet and hence slowing actuation speed.

The size, shape and composition of the emulsion jet are important parameters that alter the governing DEP force since the capacitance terms, mentioned in Eqn. 5 and Eqn. 6 become more complex.

Now, since the fluidic properties of covering oil (surface tension, viscosity) have minimal effect on the actuation dynamics [35], it is assumed that the transient behavior of emulsion jet can be modeled by suitably modifying the lumped capacitance terms and furthermore using a modified surface tension and viscosity expressions for the formed emulsion jet. Since it is utterly difficult to precisely delineate the boundary of the oil cover and hence the exact composition of the emulsion jet, an effective capacitance term (c_{eff}) was introduced to account for the lumped capacitance of the emulsion jet.

The term c_{eff} is incorporated in Eqn. 5 and Eqn. 6 and the total system capacitance is expressed as;

$$C(z) = \frac{z}{\frac{1}{c_{eff}} + \frac{2}{c_d}} + \frac{L - z}{\frac{1}{c_{air}} + \frac{2}{c_d}} + C_{drop} \tag{26}$$

c_{eff} is the effective capacitance per unit length for the aqueous-in-oil emulsion jet used to account for the lumped capacitance of the emulsion jet. The modified expression for F_{DEP} was recalculated using the principle of virtual work and after regrouping the constants it can be written as,

$$F_{DEP} = \frac{V^2}{2}\left(\frac{1}{\frac{2}{c_d} + \frac{1}{c_{eff}}} - k_1\right) \tag{27}$$

$\mathrm{k}_1 = \left(\frac{1}{c_{air}} + \frac{2}{c_d}\right)^{-1}$ is the capacitance per unit length for the part of electrode uncovered by the hemi-cylindrical emulsion jet. The term effective capacitance (C_{eff}) is essentially used to point out the alteration in F_{DEP} resulting from capacitive coupling of emulsion liquid jet. For a homogeneous jet, effective capacitance (c_{eff}) simply reduces to $k_w c_{air}$ and Eqn. 16 reduces to Eqn. 8, whereas, for an emulsion jet, c_{eff} is controlled by the composition, shape and dielectric constants of the constituting jets (oil and aqueous solution).

By suitably incorporating c_{eff} value extracted from the experimental data plots for various emulsion jets, we were able to model the modified dynamics of various emulsion liquid jets (see Table 2).

This required a curve-fitting exercise where experimental data set was fitted on to the theoretical expression in order to determine the suitability of the lumped model. Results of the curve-fitting exercise is reported in Figure 10 and a comparison of the extracted fluidic properties (labelled as theoretical fluidic properties) is presented in Table 2 to confirm that the effective capacitance model suitably predicts the behavior of different actuated emulsion jets.

Table 2. Experimental vs. extracted theoretical fluidic properties for different emulsion jets

Sample name	Expt. Jet Capacitance (C_{eff} in pF)*	Th. Jet Capacitance (C_{th} in pF)	Expt. viscosity (cSt)*	Th. viscosity (cSt)	Expt. surface Tension (dyne/cm)*	Th. surface Tension (dyne/cm)
Sample1-5cSt oil emulsion	333.54	339.08	1	2	73	72
Sample1-10cSt oil emulsion	330.78	332.16	1	8	73	71
Sample2-5cSt oil emulsion	307.25	311.4	2	4	71	69
Sample2-10cSt oil emulsion	302.40	304.48	2	8	71	69
Sample3-5cSt oil emulsion	297.56	304.48	4	7	68	66
Sample3-10cSt oil emulsion	294.79	297.56	4	9	68	67

* Jet capacitance, viscosity and surface tension values referred from [35].

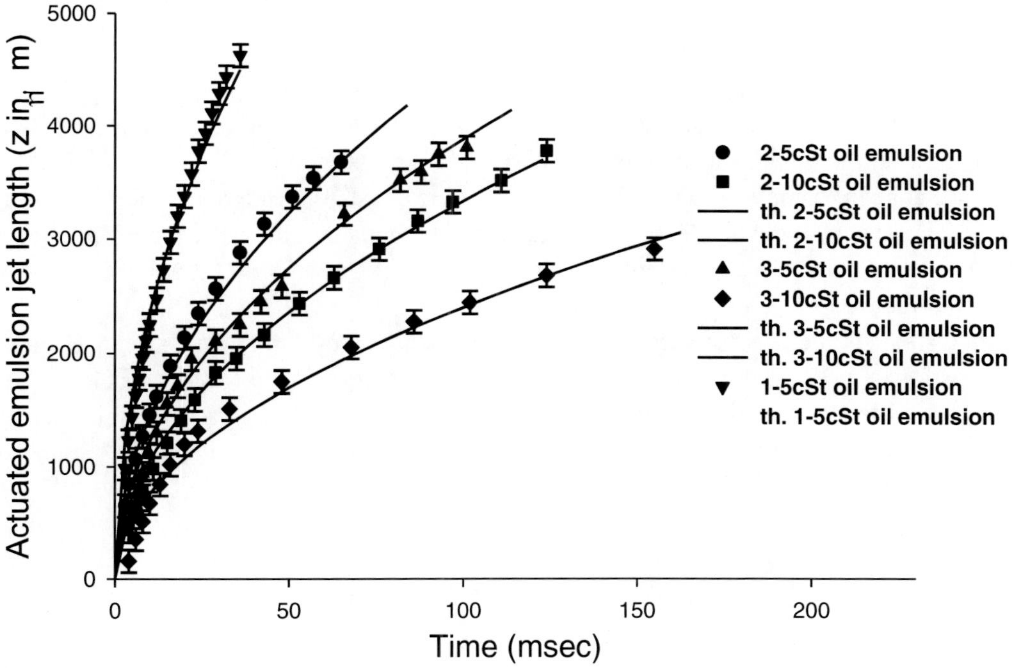

Figure 10. Results of the curve-fitting exercise based on the effective capacitance model.

2.2. L-DEP – Jet Actuation: Characteristic Features

Threshold actuation voltage

In the L-DEP actuation process, the DEP force is resisted by the fluidic surface tension and the viscous force. Of these two forces, the viscous force only becomes non-zero once the jet is initiated and covers a non-zero projection surface area (i.e. $F_\mu = 0$ at t = 0 s).

The threshold actuation voltage can be defined as the minimum actuation voltage required to initiate a liquid jet from a parent sample droplet (irrespective of whether the jet ceases to propagate after the actuation).

Based on this definition, the minimum actuation voltage can be obtained by equating the resulting DEP force with the fluidic surface tension force that tends to resist the increase in the fluidic surface area and hence the fluidic surface energy. The expression for threshold actuation voltage can be evaluated as:

$$\frac{(K_w - 1)c_d^2 c_{air} V_{th}{}^2}{2\left(c_d + 2K_w c_{air}\right)(c_d + 2c_{air})} = \pi R\gamma$$
$$\Rightarrow V_{th} = \sqrt{\frac{2\left(c_d + 2K_w c_{air}\right)(c_d + 2c_{air})\pi R\gamma}{(K_w - 1)c_d^2 c_{air}}} \qquad (28)$$

From the expression, it is clear that the minimum actuation voltage depends upon the fluidic dielectric constant, fluidic surface tension value, the electrode geometry and thickness and dielectric constant of the insulating dielectric layer, deposited on top of the metal electrodes.

Threshold actuation frequency

The frequency dependence of L-DEP actuation is due to a capacitive coupling between the energized electrode pair and the aqueous sample through the thin dielectric layer (Si_3N_4) which has a very low but finite conductivity.

Since aqueous samples have a fairly high dielectric constant, insulation on top is required in order to prevent conductive losses. To understand the impact of the frequency of applied ac voltage, we imply to the equivalent RC circuit model of the L-DEP actuation scheme, as shown in Figure 3.

In this model, the two series capacitors, C_d, represent the insulating dielectric layer on top of the two electrodes whereas, C_w and G_w are attributed respectively to the capacitance and ohmic conductance contributed by the aqueous sample. Under the assumption that the aqueous jet covers the entire electrode area (with radius R = w + g/2); the three parameters can be expressed as [27];

$$C_d \approx \frac{\varepsilon w L}{d}; C_w \approx \frac{\varepsilon_w K(1-k^2)L}{2K(k^2)} \text{ and } G_w \approx \frac{\sigma_w C_w}{\varepsilon_w} \qquad (29)$$

L is the total length of the L-DEP electrode pair, ε_w is the electrical permittivity of the aqueous sample and σ_w is the ohmic conductivity of the sample. At low frequencies, i.e. $f \ll f_c$, aqueous sample due to its high conductivity acts as an equi-potential body and virtually the entire voltage drop occurs within the thin dielectric layer. As a result, there is no electric field and hence an absence of a net DEP force.

However, even at these low frequencies, the applied field can act to manipulate the interfacial boundary of the aqueous sample-dielectric surface and hence alter the shape of the droplet and furthermore with tailored electrode geometries even achieve further droplet manipulations (splitting, transport, mixing etc…). Such low frequency actuation regime has been successfully demonstrated in EW or EWOD based SMF devices.

However, to implement DEP forces, one requires fairly high actuation frequencies ($f \gg f_c$), such that a significant portion (V_w) of the total applied voltage V can be transferred onto the aqueous droplet, sufficient enough to generate a DEP force that can overcome the fluidic resistance. This voltage fraction (V_w) along with the threshold frequency (f_c) can be analyzed from the circuit as [27];

$$\frac{Vw}{V} = \frac{C_d/2}{C_d/2 + C_w}; \qquad (30)$$

It can be demonstrated that for a typical L-DEP actuation scheme threshold frequencies are in the range of 30-50 kHz. From this exercise, one can clearly see that the percentage of applied voltage transferred to the fluidic sample and the threshold actuation frequency depend upon the dielectric properties of the fluidic sample. It furthermore indicates that for conductive biological samples (such as blood or urine samples with high ion/trace element content), or high conductivity buffers would require fairly high threshold frequency as: $f_c \propto G_w \propto \sigma_w$ and the values can go very high (in the MHz range for sample conductivity values > 100 $\mu S/cm$).

Droplet formation – key requirements/constraints

The next step during an L-DEP actuation is the break-up of the formed liquid jet into an array of droplets, termed as daughter droplets (Figure 11). The hydrodynamic instability arising from loss of equilibrium upon removal of applied AC voltage causes disintegration of liquid jet into smaller fragments. The process is governed by Rayleigh's instability criteria, proposed by Lord Rayleigh in 1879 [39]. According to Rayleigh's instability criteria, a hydrodynamically unstable liquid jet contains points/regions of maximum instability along its length, separated by a periodic wavelength (λ) which solely depends on the size and shape of the liquid jet.

The hydrodynamic instability arises once the applied voltage and the resulting DEP force is removed. The unstable jet develops constrictions along the regions of maximum instability which finally results in small cylindrical fragments on length equal to Rayleigh's wavelength: $\lambda = 9R$, where R is the radius of the cylindrical jet. Once formed, these fragments reshape in order to minimize their surface energy/surface area and recollect in form of fine daughter droplets. Size and shape of these daughter droplets can be predicted from that of the liquid jet. Jones et al [27] have demonstrated that the concept of Rayleigh's instability successfully applies in L-DEP actuation and break-up of a uniform, nearly hemi-cylindrical jet, formed during L-DEP actuation, results in formation of equi-volume and equi-spaced daughter droplet arrays.

Although, since the break-up involves retraction and recollection of fluidic mass on patterned surface, the revised instability criteria must depend on localized surface properties such as roughness, wettability, cleanliness and surface chemistry of the top layer and localized irregularities may result in irregularities in droplet dispensing. Semi-circular bumps, provided along the length of the electrode pair, may act as droplet collection sites for dispensed daughter droplets and aid to minimize localized effects [27, 29].

Furthermore, to prevent the dispensed droplets for evaporating, all L-DEP actuations like most of the other SMF schemes are carried out in a low viscosity (5 cSt) silicone oil bath. Alternatively, as described earlier, L-DPE based emulsion dispensing scheme can be utilized to dispense SE droplets by actuation and disintegrating a SE (water-in-oil) liquid jet. In this case jet break-up is anticipated to be influenced by the hydrodynamic instabilities of both the constituting liquid jets (water and oil). We observe in Figure 12 that break-up of oil sheath occurs after the inside liquid jet has already broken into daughter droplets at the specific bumps. Break-up of the oil sheath occurs over a longer time frame (10-15 msec;

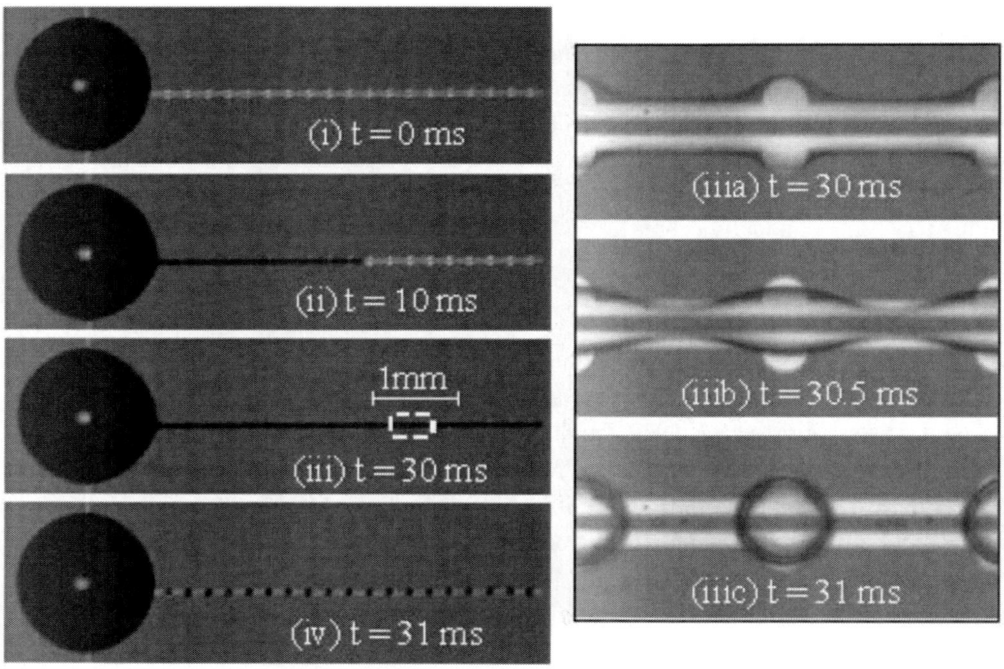

Figure 11. Micrographs showing break-up of a uniform, homogenous aqueous sample jet; (i-iv) L-DEP actuation: formation and break-up of aqueous jet; dispensed daughter droplet array; (iii a-c) close-up of jet break-up (observed under 20X magnification). Frames extracted from high speed video, captured at 2000 fps.

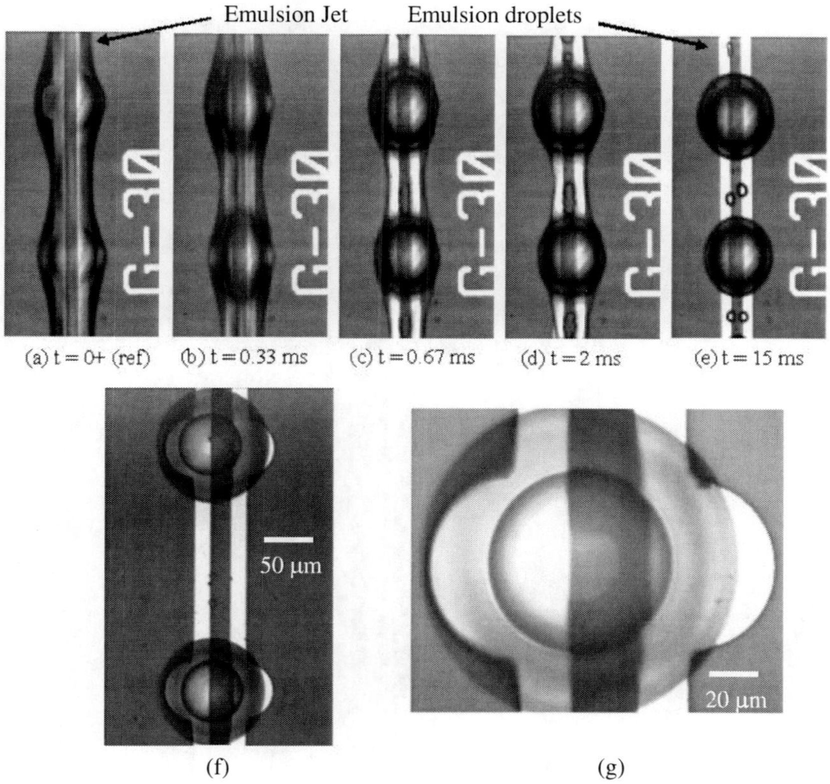

Figure 12. Video frame sequence showing breakup of aqueous-5cSt silicon oil emulsion jet and the formation of 100 picoliter emulsion daughter droplets upon removal of the L-DEP actuation voltage at frame (b). (f) Bright field images of a pair of SE droplets dispensed using L-DEP actuation scheme on w=g=20 μm electrode structure under 20x magnification; (g) a SE droplet under 50X magnification.

frames (b-e)) as compared to break-up of an aqueous jet, which occurs in 1-2 msec and it is facilitated by the instability produced as a result of formation sample daughter droplets inside the oil sheath. The dispensed droplets, shown in Figure 12 (f, g), are approximately 100 picoliter, encapsulated by a 150 μm diameter oil cover, SE droplets. To emphasize on the stability of these individually encapsulated emulsion droplets, they were observed and measured roughly up to an hour after dispensing.

Break-Up of Non-Uniform, Tapered Liquid Jets

In this section, Rayleigh's instability is applied to explain the break-up of tapered liquid jets formed over non-uniform, continuous tapered L-DEP electrodes (see Figure 5).

For a tapered liquid jet, Rayleigh's wavelength (λ) will not be a constant but rather a continuously varying parameter which decreases as the electrodes converge.

Figure 13(a-c) demonstrate the break-up of a continuous tapered liquid jet formed over a tapered L-DEP electrode pair with no form of periodic structural non-uniformity, and the resulting array of dispensed daughter droplets. We observe that the jet break-up is not periodic and the non-uniform spacing is similar to the case of irregular uniform jet break-up without bumps [27].

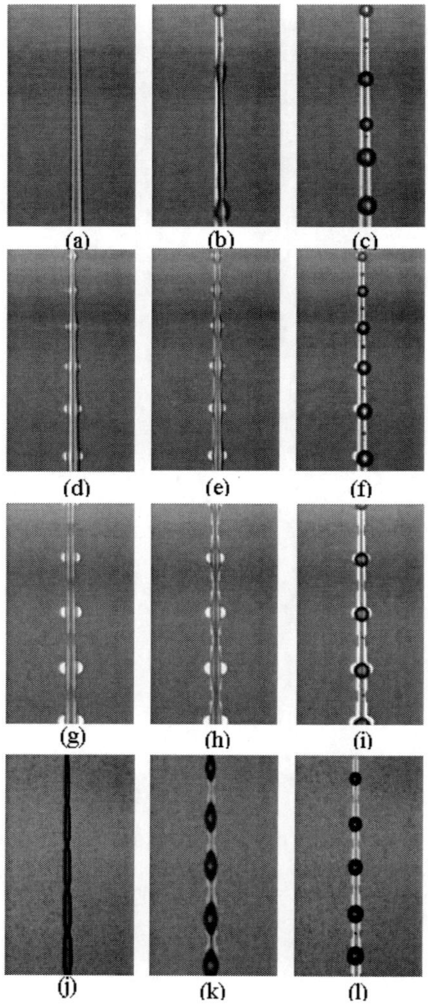

Figure 13. Micrographs showing shapes of stable DI water jet, breaking jet and the dispensed array of daughter droplets for the different continuous tapered electrode designs. (a) continuous tapered DI water jet; (b) non-reliable break-up of the continuous tapered jet; (c) dispensed non-uniform daughter droplet array; (d) tapered DI water jet on continuous tapered electrode with weighted bump spacing (weighted λ); (e) reliable break-up of the tapered jet; (f) dispensed and precisely positioned daughter droplet array; (g) tapered DI water jet on continuous tapered electrode with fixed, independent bump spacing of 500 μm and suitably incorporated pinches; (h) break-up of the tapered jet and; (i) dispensed precision daughter droplet array; (j) tapered DI water jet on pinched tapered electrode (no bumps); (k) controlled jet break-up; (l) dispensed variable volume daughter droplet array.

Figure 13(d-f) illustrate the break-up of a continuous tapered electrodes with bumps incorporated at specific bump separations that continuously decreases along the electrode length in accordance with the Rayleigh's wavelength. It is evident that the placement of dispensed daughter droplet is more precise and repeatable as compared to the case without bumps. Although the bumps play no active role in jet break-up and are only incorporated to overcome any localized irregularities during recollection of the disintegrated jet to form daughter droplets, they do create small distortions in the shape of the formed liquid jets. The distortions become more prominent when the liquid surface tension is reduced (by adding surfactants or other variants) [35].

Since bumps do not actively participate in the actuation and jet break-up process, using an alternative scheme to induce electromechanical instability along the L-DEP electrodes appears to be more favorable. Such an electromechanical instability can be introduced in form of periodic constrictions (pinch) along the electrode length (see Figure 5(c)) which can actively control the electrical and fluid mechanical properties of the jet during actuation and break-up by altering the shape and boundary conditions for the tapered liquid jet [36]. When suitably incorporated these pinched regions become specific regions/points of maximum instability, allowing the user to modify λ and thus create an array of daughter droplets with controllable droplet volume and spacing. An example of pinched electrode geometry is shown in Figure 5(c) where the pinched regions are positioned separated by an arbitrary length L which does not correspond to λ of any specific section of the formed tapered jet. Pinched scheme with or without bumps were observed during L-DEP actuation. Break-up of these pinched tapered jets and the dispensed droplet arrays are shown in Figure 13(g-l). The observation clearly suggested that a pinched electrode scheme is more suitable for reliable jet break-up and droplet array dispensing and furthermore such a scheme is less susceptible to changes in fluidic properties, especially surface tension, affecting the break-up.

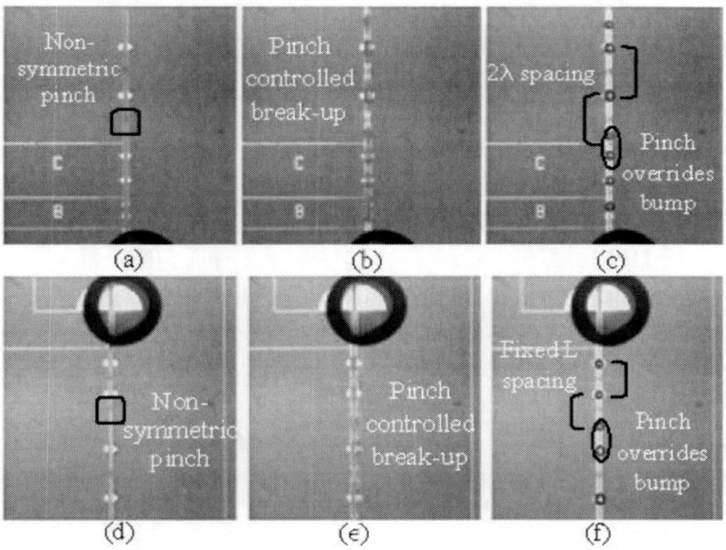

Figure 14. Micrograph images showing actuation and droplet dispensing over a uniform (w = g = 30 μm) L-DEP electrode with various arrangements of bumps and pinches for a comparative analysis of their contribution in controlling the jet break-up. (a) Actuated DI water jet over one end section of L-DEP electrode containing λ/4, λ and 2λ bump separations and specifically indicating placement of one asymmetric pinch within the first 2λ part; (b) Break-up of the modified jet which is strongly controlled by the position of various pinches; (c) Precision droplet formed at 2λ spacing but not in λ/4 spacing showing the override in droplet positioning due to the asymmetrically placed pinch (between the λ and 2λ regions); (d) actuated DI water jet over the alternate end section of L-DEP electrode scheme containing fixed L and 2λ bump separations and specifically indicating placement of one asymmetric pinch between the fixed L and the 2λ part; (e) Break-up of the modified jet and (f) Precision droplet formed at 2λ and fixed L spacing whereas also showing the override in droplet positioning due to the asymmetrically placed pinch (between the fixed L and 2λ regions).

Another interesting scenario observed was to compare the effect of 'pinch' to the bumps in controlling jet breakup. To facilitate such a comparison, a different uniform electrode scheme (shown in Figure 14(a)) with four different sections of bump separation $\lambda/4$, λ, 2λ and an arbitrary separation length L ($L > \lambda \neq n\lambda$), was used. Some of the sections on this L-DEP electrode arrangement consist of bumps placed at λ but pinched in between in a non-symmetric fashion (see Figure 14). As a result, upon jet break-up, the daughter droplets are not formed at the bumps separated by λbut symmetrically around the pinched region (see Figure 14(c, f)). On another section of this scheme, we have placed very small sized pinched region, compared to λ, and in this case upon break-up, positioning of the daughter droplets is not controlled by the pinch. This confirms that a suitable designed pinch can partially or successfully alter Rayleigh's wavelength and in turn allow the user to program and control the droplet dispensing process.

2.3. Droplet-Dielectrophoresis (D-DEP): Underlying Concept and Actuation Methodology

Similar to L-DEP, D-DEP actuation requires a pair of coplanar metal electrodes patterned on an insulated substrate. A typical D-DEP actuation scheme and the mechanism of droplet actuation are shown in Figure 15. The coplanar metal electrodes are diagonal fishbone shaped and are again insulated on top with a thin insulating layer (Si_3N_4) and are rendered hydrophobic (refer device fabrication section) using Teflon®. The actuation mechanism is a combination of the two phenomena namely, DEP and EW. When the electrode pair is energized with a low voltage (~ 60-120 Vpp) and low frequency (30-100 Hz) AC voltage, it induces periodic deformations in the droplet which is placed at one end of the D-DEP electrode (see Figure 15(b)).

It can be observed from the geometry that the resulting electric field will be very intense near the edges of the fishbone shaped electrodes which results in the meniscus of the droplet being dielectrophoretically pinned at the edge.

The resulting deformation is a unidirectional, non-symmetrical one (Figure 15), causing a net shift in the centre of mass (CM) of the droplet. When the E-field is periodically turned off (obvious during an AC voltage application), the pinned trailing edge of the droplet retracts, resulting in a net forward motion of the entire droplet.

This periodic deformation and shifting takes place during the entire actuation time, resulting in the droplet being successfully transported from one end of the electrode to the other end. Figure 15(b) shows periodic droplet motion during a typical D-DEP actuation.

The frequency of actuation used for D-DEP actuation is restricted by the fluidic response (and hence the fluidic damping) to the applied AC frequency. It was observed that at high frequencies (> 150 Hz), an aqueous droplet is unable to resonate with the applied frequency, resulting in mere oscillations of the droplet while pinned at one specific point. In case of D-DEP actuation under oil bath, the actuation is further lowered due to viscous damping provided by the surrounding oil media. Successful D- DEP actuations have been achieved at lower frequencies (~ 50 Hz) while actuating droplets under low viscosity (5-10 cSt) silicone oil bath.

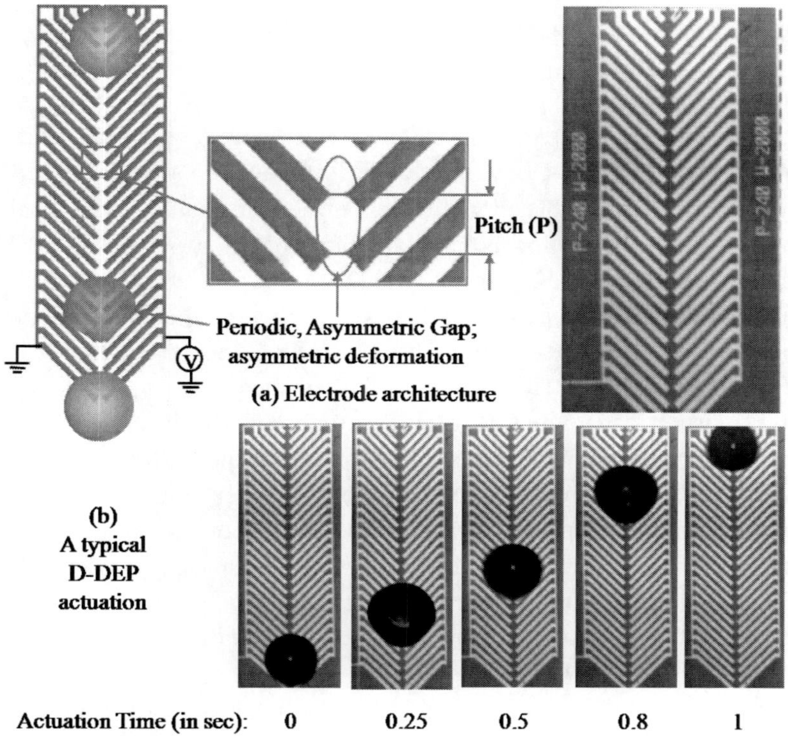

Figure 15. a) D-DEP electrode architecture; (b) Droplet transport during a typical D-DEP actuation process.

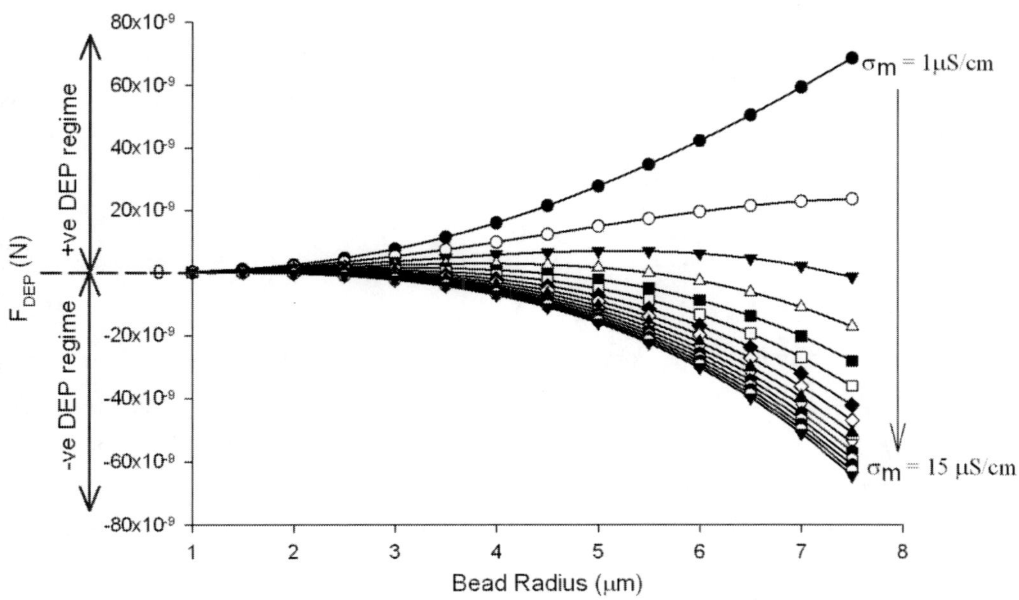

Figure 16. Plots showing variation of DEP force for different sized micro-beads actuated in varying conductivity fluidic media at 100 kHz.

2.4. Manipulating Micro-Particles Leveraging L-DEP

We mentioned earlier that DEP-LOC schemes can be utilized for size and dielectric property based screening of living/dead cells, micro-sized polystyrene beads and other micro-particles, also termed as particle-DEP. Jones et al. [40] have reported that when polystyrene micro-beads (dia. < 1 μm) are actuated within de-ionized water sample over a pair of uniform L-DEP electrodes, their distribution is largely influenced by a positive DEP force on the beads which affects their effective transport during the actuation.

Particle-DEP offers flexible and controlled segregation methodology as the nature and effects of the DEP force can be altered by controlling the conductivity and dielectric properties of the carrier fluid. However, for this segment, our focus is to look at the dispensing and transport capabilities of L-DEP actuation since, in order to achieve 'bead-based' bio-detection and screening assays, L-DEP devices should be capable of dispensing ultralow quantities of bio-samples and micro-beads; either uniformly or of controlled volumes, across the array of dispensed droplets. The work reported here was conducted under specific fluidic conditions to establish L-DEP as a tool for rapid screening of functionalized micro-beads by optimization of the fluidic medium properties.

During L-DEP actuation of the liquid jet, the suspended micro-particles are subjected to a DEP force orthogonal to the flow, as a result of the non-uniform electric field emanating from the energized micro-electrodes. The DEP force acting on a spherical particle is expressed as [40]:

$$\vec{F}_{DEP} = 2\pi\varepsilon_m R^3 \operatorname{Re}\left[\bar{K}\right]\vec{\nabla}\left(\left|E_{rms}\right|^2\right) \tag{31}$$

where R is the particle radius, ε_m is the medium permittivity, E is the applied electric field and $\bar{K}$ is the frequency and material/media property dependent Clausius-Mossotti factor given by;

$$\bar{K}(\omega) = \frac{\bar{\varepsilon}_p - \bar{\varepsilon}_m}{\bar{\varepsilon}_p + 2\bar{\varepsilon}_m}, \quad \bar{\varepsilon} = \varepsilon - j\,\sigma/\omega \tag{32}$$

The quantity $\bar{\varepsilon}$ represents the complex permittivity of the particle and that of the liquid media, ω is the frequency of the applied electric field and σ is the electrical conductivity.

The sign of $\operatorname{Re}(\bar{K})$ determines the nature of the DEP force and its value lies in the range [-0.5 to 1.0] for the entire frequency spectrum.

The four parameters: ε_p, σ_p, ε_m, σ_m, collectively modulate the DEP force, resulting in positive or negative DEP of particles in different regions of the frequency spectrum and a threshold frequency where it switches sign/direction. This cross-over frequency (ω_c) is given as [41]:

$$\omega_c = \sqrt{\frac{\left|\left(\sigma_m - \sigma_p\right)\right|\left(\sigma_p + 2\sigma_m\right)}{\left(\left|\varepsilon_p - \varepsilon_m\right|\right)\left(\varepsilon_p + 2\varepsilon_m\right)}} \quad (33)$$

Thus based on this simplified particle model, there are two regimes of DEP forces that can influence the spherical particle during L-DEP actuation, separated by a cross-over frequency (ω_c). The Clausius-Mossotti factor, for frequencies sufficiently lower or higher than the corresponding ω_c, is dependent on the conductivity and permittivity differences between the particle and the suspension medium and can be simplified as:

At low frequencies $(\omega << \omega_c)$: $$\bar{K}_0 = \frac{\sigma_p - \sigma_m}{\sigma_p + 2\sigma_m} \quad (34)$$

At high frequencies $(\omega >> \omega_c)$: $$\bar{K}_\infty = \frac{\varepsilon_p - \varepsilon_m}{\varepsilon_p + 2\varepsilon_m} \quad (35)$$

The conductivity of the micro-particle is difficult to estimate, as it depends on both the bulk conductivity (σ_b) and the surface conductance (K_s) of the particle whose respective contributions are as follows:

$$\sigma_p = \sigma_b + \frac{2(K_s + K_i)}{R} \quad (36)$$

K_i is the conductance term added due to the ionic, diffused double layer. In case of standard latex beads with a polystyrene core, the bulk conductivity is negligible (σ_b ~ 0) whereas, K_s is calculated and reported close to 1.2 nS (referred from [42]). In low ionic media (such as glycerol, very low conc. salt soln.), term K_i can be neglected when estimating the particle DEP.The DEP force, assuming that the non-uniform electric field inside the aqueous jet is nearly azimuthal, is expressed as:

$$\bar{E}_\theta(r) \approx {}^{V_a}\!/_{\pi r}(\hat{\theta}) \quad (37)$$

Since the electric field is capacitive coupled, voltage across the liquid finger is represented by;

$$V_f \approx \frac{C_d}{2C_m + C_d} V_a \quad (38)$$

where, V_a is the applied AC voltage. The expressions for the capacitance terms, for both uniform and tapered liquid jet actuation, have previously been reported in Eqn. 5, 6. Using Eqn. 30 and Eqn. 36, the DEP force can be determined to be [40];

$$\overline{F}_{DEP,r} = -\frac{4\varepsilon_m R^3 \operatorname{Re}[\overline{K}]}{\pi r^3} V_f^2 \left(\hat{r}\right) \tag{39}$$

In order to avoid the regime of positive DEP forces for the used micro-particles, which can be estimated using Eqn. 38, at 100 kHz (the actuation frequency), it is crucial to suitably adjust the media conductivity such that $\sigma_m > \sigma_p$; and the DEP force remains negative. Under these conditions, the micro-beads are pushed up and away from the strong field regions near the electrode surfaces. We used DI water-glycerol mixtures to alter the rheological and electrical properties of the fluid media (see Table 1) and observe actuation of polystyrene micro-beads under those conditions.

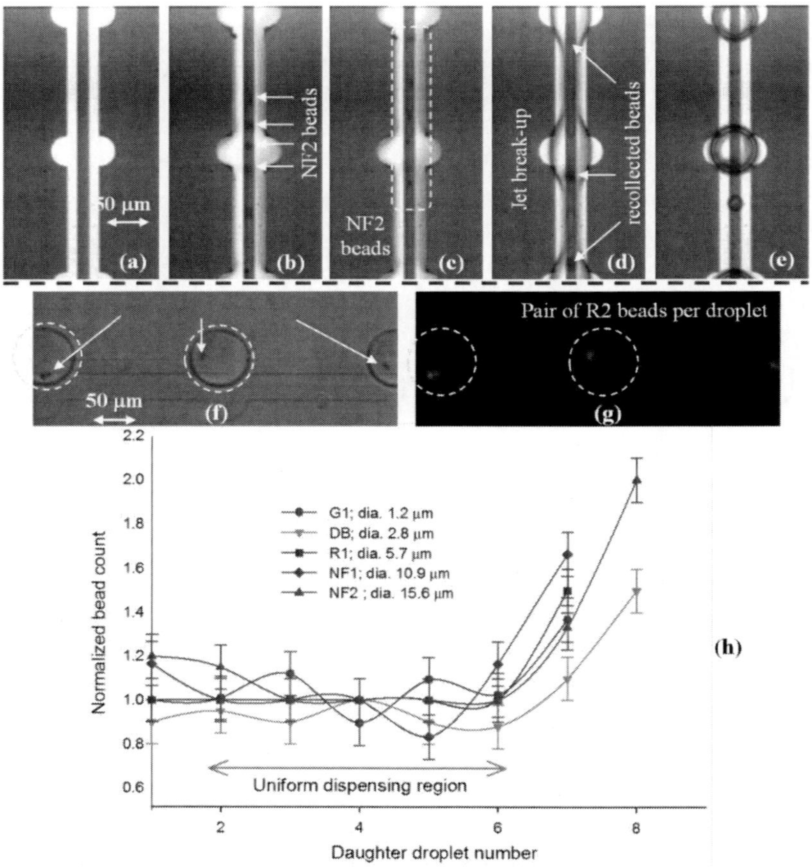

Figure 17. Micrograph collage showing different concentrations and sizes of microbeads actuated and dispensed using LDEP. (a–e) Transport and dispensing of polystyrene microbead (NF2: dia. – 15 µm) during LDEP actuation on w = g = 20 µm LDEP electrodes; (f, g) bright field and fluorescent images of three adjacent droplets (~1 nL) containing pair of R2 beads during LDEP actuation process; (h) plots showing distribution of various micro-beads, actuated over a w=g=20 µm L-DEP electrode scheme.

Increasing the glycerol concentration causes increase in medium density and viscosity while the relative dielectric constant of the mixture decreases and the medium ionic conductivity increases but stays in the range of (2.5 - 15 μS/cm).

Also, glycerol is a suitable media for bio-applications as the osmotic pressure of such mixtures are closely comparable to that of cellular fluids and thus biocompatible.

Numerical analysis was conducted to determine the range of variance of F_{DEP} and cross-over frequency for the different fluidic media (Table 1) and five different bead samples (bead diameter: 1.2 μm to 15 μm) under L-DEP actuation (AC frequency = 100 kHz and actuation voltage = 450 V over w=g=20μm L-DEP electrode) and presented in Figure 16.

The analysis suggests that medium conductivity of nearly 8-15 μS/cm is required in order to avoid the regime of positive DEP for the used bead samples. For cross-over frequency, it was established that the L-DEP actuation frequency (100 kHz) is lower than the cross-over frequency for the entire experimental regime (data points in Figure 16), hence justifying the use of Eqn. 34 in the numerical model.

This numerical analysis and its experimental implementation confirms that the effective force acting on the particles during such L-DEP actuations will result in an upward trajectory for the micro-beads within the transient flow. The numerical analysis was further backed by results showing uniform and controlled distribution of the different micro-bead samples during L-DEP actuation process as reported in Figure 17.

3. Practical L-DEP Devices- Fabrication Methods

3.1. Fundamentals of Device Fabrication

L-DEP microfluidic devices are fabricated in clean-room environments, using conventional micro-fabrication processes. The morphology of a typical L-DEP SMF device as eluded to in theory section, consists of a pair of micro-electrodes, patterned in metals such as Aluminum or, transparent metal oxides such as ITO (Indium Tin Oxide), used to generate sufficient spatially non-uniform electric field, required for liquid actuation. The microelectrodes serve as open fluidic tracks over which liquid samples are dispensed in form of microscopic sized droplets and further conveyed (to different on-chip reaction site), mixed (with other sample/reagent droplets) and manipulated (splitting, separation etc…). However, since most actuated liquids are aqueous samples with fairly high conductivity values, it is necessary to insulate the micro-electrodes with a thin but sufficient dielectric layer (e.g. silicon nitride, silicon oxide, PDMS), in order to prevent sample electrolysis during actuation. For large scale integrated electrode schemes, multiple electrode layers can be patterned with sandwiched dielectric layers to create multiplexed electrode schemes. The top surface of the device is then rendered hydrophobic by adding a thin layer of any hydrophobic polymeric materials (typically Teflon®). A hydrophobic surface provides a high contact angle which facilitates droplet dispensing and further manipulations. Based on the discussed structural necessities, we now present a typical clean-room fabrication process to fabricate a surface microfluidic device on Silicon substrate:

a. Fabrication process starts with substrate cleaning with piranha solution (H_2SO_4:H_2O_2::3:1) for roughly 20 minutes.
b. A thin metal layer (150-200 nm) is deposited onto the substrate using magnetron sputtering. Metals such as Al, Au, Cr are suitable and can be easily patterned whereas transparent metal oxides such as ITO (Indium Tin Oxide) are also useful due to their transparency and low sheet resistance [43, 44].
c. Positive photoresist (HPR504) is spin-coated on top of the metal layer and soft baked at 110°C for ~ 90 sec and then hydrated for around 10 minutes.
d. A standard photolithography unit with mask aligner and UV exposure lamp is used to de-polymerize the exposed photoresist. The exposed photoresist is removed using a developer solution (Developer 354), leaving exposed metal layers.
e. Wet chemical etching is then used to remove the exposed metal layer, resulting in generation of patterned metal electrodes, covered with undeveloped/unexposed photoresist. The left-over photoresist is washed off using acetone/IPA.
f. To passivate/insulate the micro-electrodes, a thin layer (~300-600 nm) of dielectric material (Si_3N_4 or, SiO_2) is deposited using plasma enhanced chemical vapor deposition (PECVD) technique.
g. A second layer of photoresist is then spin coated, exposed to UV light, developed and patterned as explained above.
h. The exposed insulating layer (over the bonding pads) is etched using a reactive ion etching.

If a second metal layer is to be deposited (for two layered multiplexed electrode schemes), steps b. to step h. are repeated in order to generate the dual metal layers insulated by the sandwiched dielectric layer. The final top surface of the SMF chip is rendered hydrophobic and this could be achieved in many a ways, the simplest being spin coating of a thin layer (~ 100 nm) of Teflon® AF Resin (DuPont Inc., USA).

3.2. Approaches for Reliable and Repeatable Device Operation

There are several crucial aspects to consider while fabricating a reliable DEP-LOC device but none are more important than the insulating dielectric layer and its top surface. An ideal insulating material should possess: chemically inert surface for reproducible LOC applications, fairly high contact angles (~120^0 or more), high dielectric strength or breakdown field strength, ability to be wielded in thin layers and minimal material defects. To this date, no single material can satisfy all the above mentioned features and as a result, several combinations of dielectric/insulating materials are used to provide the desired characteristics of the device.

For example: we use Si_3N_4 as the thin insulating layer since it has very high breakdown strength but poor top surface properties (very low water contact angle). So, it is used in combination with a thin hydrophobic top coating (e.g. Teflon AF®) to overcome this limitation [17]. Based on the applied high voltages and the break-down strength of Si_3N_4, the layer thickness should be kept in range of 250-500 nm whereas ~ 50 nm of Teflon is found sufficient to achieve successful L-DEP actuations. However, Teflon suffers from poor adhesion onto Si_3N_4 layer and limitations have been found for this combination of dielectric stack for repeated or multiple actuations [45]. More recently, a plasma-deposited fluorocarbon (FC) layer is incorporated prior to the spin-coated Teflon deposition to provide a more robust and reliable composite coating. To fabricate this "composite coating", the device surface following Si_3N_4 insulation is first cleaned using a short span of plasma etching.

This is followed by plasma-deposition of thin FC layer (~30-40 nm) using PECVD, inductively coupled plasma (ICP) and C_4F_8 as precursor gas [46]. Thin plasma FC films (~30 nm) are thus formed and then spin-coated to form the composite layer. The combined stack of Si_3N_4 and the composite hydrophobic layer, is proven to be ideal to fabricate reproducible DEP based SMF devices [45].

3.3. Process for L-DEP and D-DEP - Multilayer and Device Integration

So far in this chapter, we have demonstrated dispensing capabilities of different L-DEP schemes and droplet transport capability of D-DEP scheme. In order to achieve a DEP based LOC device, we must integrate the two electrode architectures to devise integrated electrode designs. The simplest of such integration, shown in Figure 18(a), is a $1x1$ integrated actuation scheme which was proposed by Chugh and Kaler [34]. A binary combinatorial assay on such a scheme takes place in a two-step process. Parent sample and reagent droplets (~ 1μL) are placed on the parent droplet sites, at the end of the two L-DEP electrode pairs.

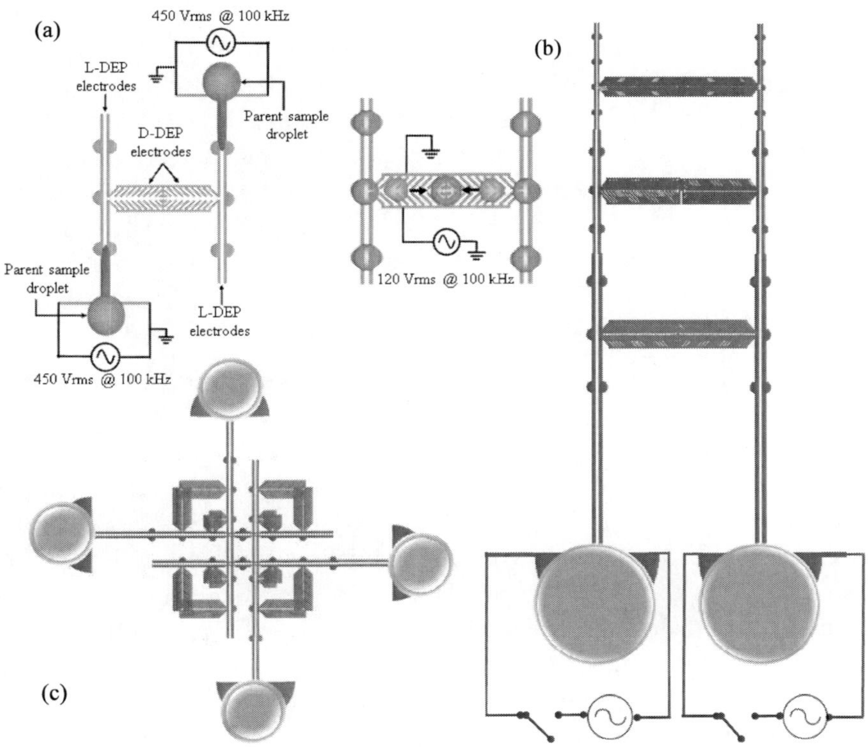

Figure 18. Schematic diagrams of (a) $1x1$ integrated uniform L-DEP D-DEP scheme; (b) $1x1$ integrated tapered L-DEP D-DEP scheme and (c) $2x2$ multiplexed L-DEP D-DEP matrix.

The L-DEP electrode pairs are simultaneously energized using a high voltage, high frequency AC signal (~ 450-500 Vpp at 100 KHz) resulting in formation of two liquid jets, each emanating from the sample and reagent droplets and rapidly propagating along the two L-DEP electrodes (Figure 18(a)). Once the jets have entirely covered the L-DEP electrode (L-DEP actuation time ~ 15-20 ms), the applied voltage is removed, resulting in break-up of the two liquid jets and formation of daughter sample and reagent droplets, placed precisely at droplet collection sites (bumps) along the two L-DEP electrodes. At this point, the fishbone shaped D-DEP electrodes are energized using a low voltage, low frequency AC signal (~ 120 Vpp at 50 Hz) resulting in specific sample and reagent daughter droplets being transported in parallel, automated fashion towards the reaction/mixing site (the merge/overlap of two adjacent D-DEP schemes) as shown in Figure 18(a).

Once the sample/reagent droplet pairs are in close vicinity, binary mixing is achieved and the voltage is kept on to enable electric field based stirring and to facilitate efficient binary mixing. Such a *1x1* integration can also be achieved between tapered L-DEP and D-DEP electrode designs (Figure 18(b) which are useful for titration based assays [47].

These simple scheme inspired further attempts to generate DEP based large scale integration systems that could dispense and furthermore manipulate multiple sample and reagents in parallel, automated fashion in order to generate multiplexed reaction chambers. A two layered fabrication process (see device fabrication) is utilized to generate L-DEP and D-DEP actuation schemes combining to produce a *2x2* multiplexed reaction chambers which were then utilized to perform multiplexed quantitative biochemical assays. A typical *2x2*, multiplexed electrode scheme is shown in Figure 18(c). It consists of two set of L-DEP electrode schemes, in two different metal layers, separated by a thin (~ 500 nm) dielectric layer sandwiched in between to avoid short-circuiting of electrodes. The D-DEP electrode schemes were integrated with the L-DEP electrode schemes in both layers, to generate a matrix of electrodes and the reaction/mixing sites were symmetrically placed in the four corners of the multiplexed matrix (Figure 18(c)). These integrated electrode schemes have been used to conduct several quantitative, combinatorial, biochemical assays which are discussed in the following section.

4. Bio-Sample Manipulation and Sensing Applications

In this section, we present applications pertaining to the various L-DEP actuation methodologies for on-chip bio-detection and DNA/RNA analysis. The opto-electronic experimental set-up used during the capturing of the presented data (graphs, bright/fluorescent images and quantitative data) is briefly explained here.Figure 19 shows a schematic description of the complete opto-electronic set-up along with screenshots showing some of the more important components of the set-up. In the set-up, the printed circuit board (PCB), consisting of multiple arrays of spring loaded pogo pins and two high voltage relay switches (9104 series, COTO Technology, USA) is used to provide the voltage supply to the SMF chip. The chip itself is secured on the PCB with the help of a Plexiglas piece, designed to hold the arrangement together (with four screws and slot for placement of the SMF chip) during experimentation (Figure 19). To facilitate rapid and automated biochemical assays and to minimize external perturbations, the entire opto-electronic set-up is controlled using an NI

(National Instrument) Labview interface. The various applications of the SMF device captured using the detailed set-up are categorized and presented in the three sections below.

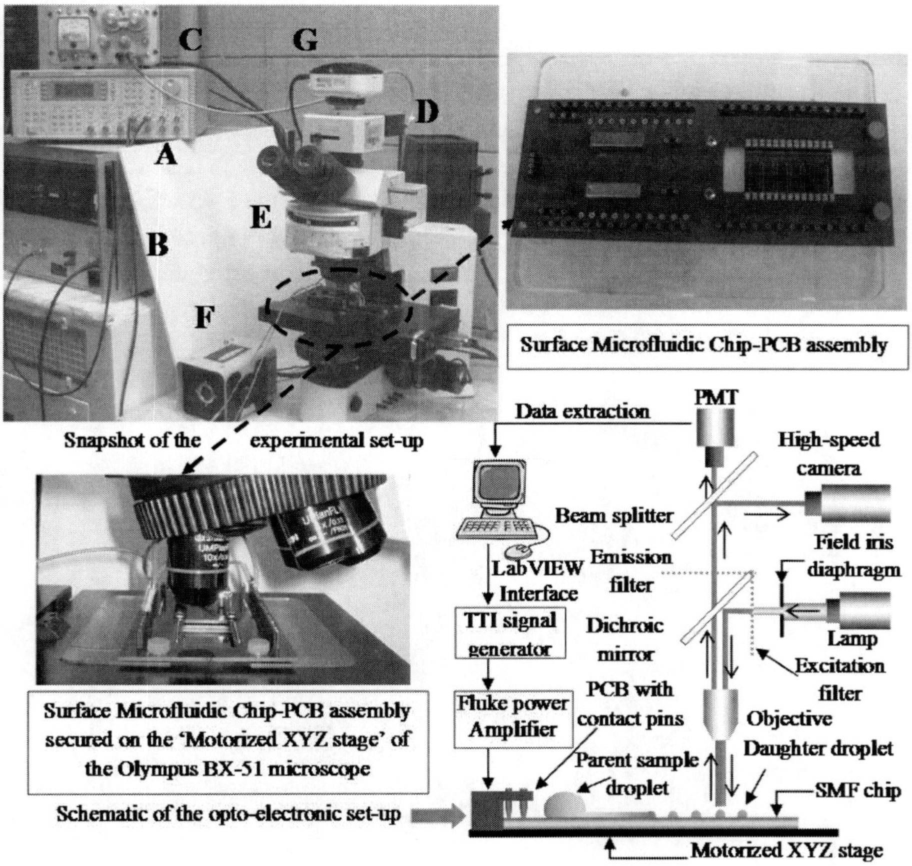

Figure 19. Snapshots and schematic diagram of the opto-electronic experimental set-up: A. Arbitrary waveform generator (TGA 1244, TTi, UK); B. Precision power amplifier (5205A, Fluke, USA); C.Precision power source (Model 2005, Power design Inc., USA) for the Photo multiplier tube; D. PMT (Photosensor module H7468, Hamamatsu, Japan); E. Fluorescence Microscope (BX51, Olympus, Canada) with a motorized XYZ platform; F. High Speed Camera (MS 70K, Canadian Photonics Inc., Canada); G. CCD camera (Micropublisher 3.3 RTV, QImaging, Canada).

Droplet based bio-assays

In this section, we report on the combinatorial bio-assays conducted using the integrated electrode schemes (shown in Figure 18) for dispensing, transport and mixing of bio-sample and reagent droplets. Detection and quantification of nucleic acids is a key aspect of life sciences research and integral to a wide range of biochemical assays such as polymerase chain reactions or nucleic acid extractions from cells [48, 49], DNA hybridization studies and quantifying DNA fragments in cDNA library production [50, 51].

For sensitive analysis, especially on a microfluidic platform where sample volumes are typically in the range of nL to pL, fluorescence based detection is the most suitable. In fluorescence based DNA analysis, nucleic acid molecules are stained with certain fluorochromes, which produce a characteristic fluorescence emission when excited by light of suitable wavelengths [52, 53].

We first report on a fluorescence based DNA quantification assay using the integrated tapered L-DEP and D-DEP actuation scheme [47], illustrated in Figure 18(b).

In this study, aliquots of dsDNA (pUC57 plasmid) and a popular nucleic acid stain, PG (PicoGreen® Molecular Probes, Invitrogen) were individually actuated on a pair of tapered L-DEP electrodes (Figure 20(a, b). The dispensed daughter droplets of both sample and reagent thus formed were subsequently actuated using the three integrated D-DEP electrode pairs and mixed (Figure 20(c-f). The tapered dispensing allows sequential reduction in sample and reagent quantities in the mixed droplets (Figure 20(g-i). The fluorescence emission from the mixed droplets (Figure 20(g-l)) was quantified using the PMT to estimate the amount of DNA present in each mixed droplet. The PMT data confirmed controlled deposition of DNA sample in the mixed daughter droplets [47].

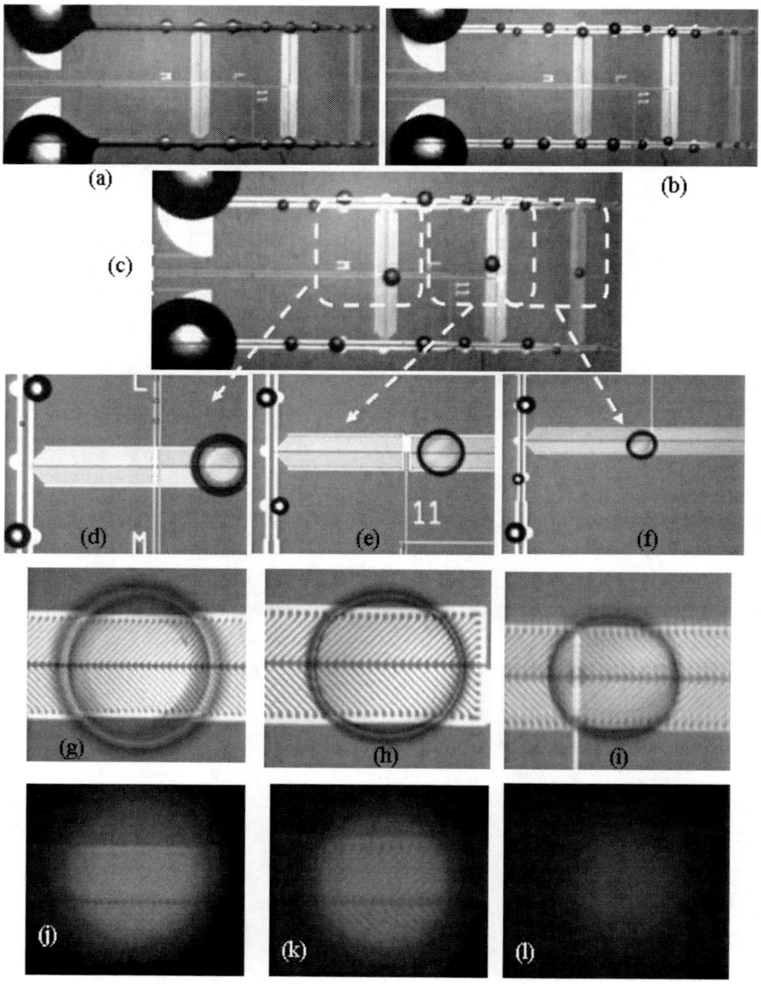

Figure 20. Micrographs showing the result of a variable volume DNA quantification assay. (a, b) tapered L-DEP actuation and droplet dispensing; (c) mixed daughter droplets; (d, e, f) Unmixed DNA sample droplets (along tapered L-DEP electrodes) and mixed DNA-PG droplets; (g-l) bright field and fluorescent images of the mixed DNA-PG droplets showing the fluorescent intensity gradient.

Table 3. Structural details for the used oligonucleotide samples

Name	5'	Sequence	3'	λ/λ*	Conc.
P1	Iowa Black™ FQ	TGATGGTGCCTATTTTACGG	-		15μM
P2	Iowa Black™ RQ	AAGCAGAGACTATTTTACTG	-		15μM
S1	-	CCGTAAAATAGGCACCATCA	6-FAM™	494/520	4μM

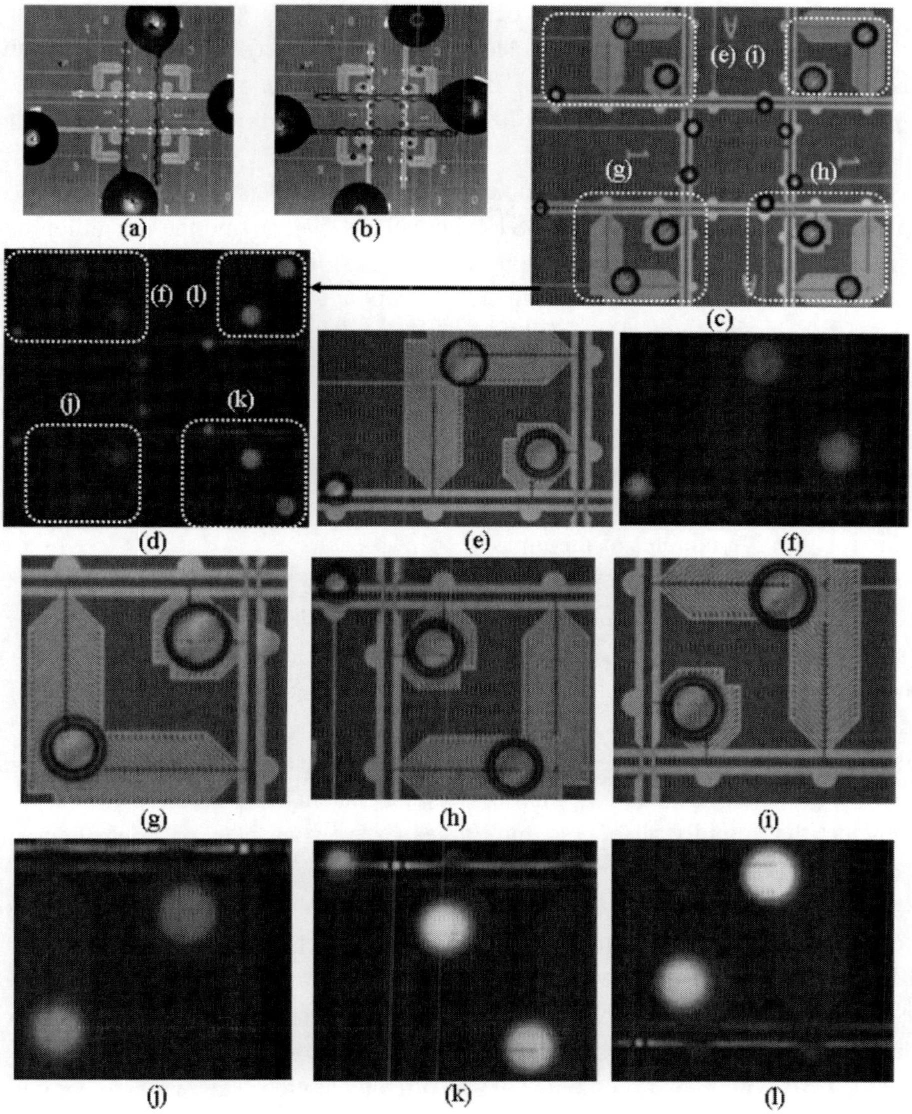

Figure 21. Results of the multiplexed DNA hybridization detection assay. (a, b) Top and bottom layer L-DEP actuation (500 msec); (c, d) Bright field and fluorescent image of the four different corners of the 2X2 matrix after D-DEP actuation (3-4 sec); (e-l) Bright field and fluorescent images of the four corners showing quenched (e, g) and unquenched mixed droplets (h, i).

In another droplet based detection assay, we present a DNA hybridization analysis on a 2*x*2 SMF device [54]. The 2*x*2 scheme consists of four sets of L-DEP electrodes arranged in a matrix like geometry along with D-DEP electrodes integrated at suitable locations (see Figure 18(C)). The 2*x*2 SMF device was used to actuate three different oligonucleotide samples, labeled with different reporter and quencher fluorochromes, and allowed us to study selective DNA hybridization detection using the principles of fluorescence resonance energy transfer method (FRET) [55, 56]. The oligonucleotides, purchased from Integrated DNA Technologies (USA) are reported in Table 3. Oligo S1 was designed as complimentary to P1 whereas P2 was designed with non-complementary base sequences.

Probe oligo samples (P1 and P2) contained a quencher molecule to selectively quench the 6-FAM fluorophore when hybridization occurs. The two probe samples are actuated on the vertical L-DEP electrodes (P1 on left electrode) (Figure 21(a)) whereas two identical S1 samples are actuated on the horizontal L-DEP electrodes (Figure 21(b)).

The matrix of D-DEP electrode is utilized to transport and mix the dispensed sample and reagent daughter droplets (Figure 21(c)).

This results in four pairs of mixed daughter droplets, two of which contain complimentary DNA sequences, P1 and S1 (Figure 21(e, f, g, j)) and the remaining two pairs contain P1 and S2 with non-complementary sequences (Figure 21(h, i, k, l)). PMT response to the fluorescence emissions of these mixed droplets and remaining unmixed droplets was measured to calculate the quenching efficiency and hence the effectiveness of the multiplexed assay. The two pairs of mixed droplets containing non-complimentary sequences show very low quenching percentage (20-28%) (Figure 21(f, j)) which results primarily from the quenching due to the adjacent nucleotide (Quenching efficiency for FAM™-Adenosine ~23%) as reported by [56]. The four mixed droplets (Figures 21(k, l)) that contain the complementary oligos P1 and S1 show a fairly high value of quenching efficiency (~70-75%) as measured within 3-5 minutes of mixing.

Vesicular bio-applications

Lipid bilayer vesicles can be spontaneously assembled, often using off-chip methods such as ultrasonication, phase inversion, electroformation and micro-channel technology [57, 58, 59].

However, most of these schemes are very limiting for on-chip applications and often require complex steps with large volume of solvents. In this section, we present an L-DEP based scheme that rapidly dispenses multitude of lipid vesicles of controlled size and structure.

Furthermore, the applicability of SMF chips in assembling variable sized giant vesicles (diameter: 10 – 100 μm) which can be leveraged to anchor and transport Quantum Dot labeled bio-probes in order to develop vesicular bio-sensors for tailored bio-detection [37, 60].The bio-samples used in this work include: two biotinylated phospholipids and two non-biotinylated phospholipid samples (Avanti Polar Lipids); Qdot®-525 streptavidin conjugate (Invitrogen, USA) and two complementary oligonucleotide samples (Integrated DNA Technologies, USA).

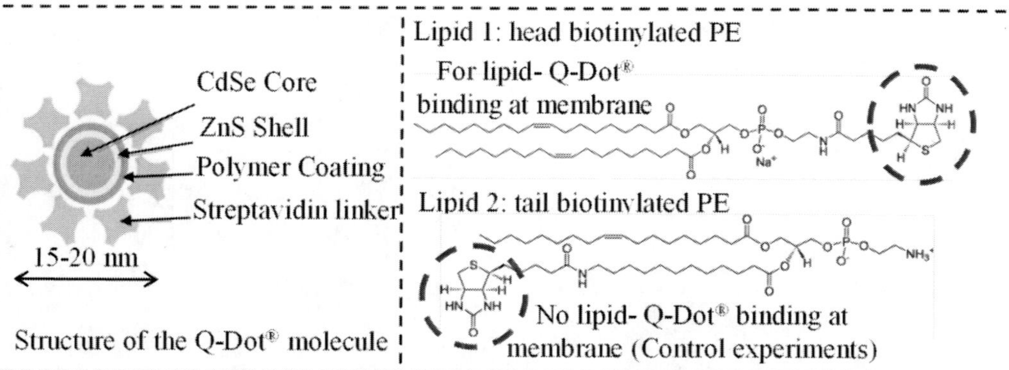

Oligonucleotide samples:
P1: 5'-IOWA FQ™ quencher; 20 bases; for Q-Dot quenching
5- /5IAbFQ/TGATGGTGC CTATTT TAC GG/3Bio/ -3
S1: 3'-biotinylated; 20 bases complementary to P1 for hybridization
5- CCGTAA AAT AGG CAC CAT CA/3Bio/ -3

Figure 22. Structural attributes of various bio- samples used in experimentation.

The oligonucleotide concentrations used during experiments lied in the range of 10-30 ng/µL. Structural details of these bio-samples are reported in Figure 22. Lipid bilayer vesicles were synthesized on-chip using the previously explained L-DEP based emulsion dispensing scheme, as elaborated in Figure 23 [35]. On application of AC voltage (480 Vpp @ 100 kHz), an emulsion jet emerged from the parent droplet and rapidly (typically within ~5-10 msec) conveyed over the electrodes. Upon removal of the electric field, jet disintegrated to form sub-nanoliter sized emulsion daughter droplets (Figure 23: step 1). The resulting precision emulsion droplets contained lipid monolayer at their water-oil interface (Figure 23). The outer cover of these emulsion droplets was then diffused in a comparable viscosity dispersion media (doped with silicone oil) bath resulting in stable lipid monolayer formation which encapsulates the aqueous solution within the silicone oil bath (Figure 23: step 2).

Finally, to assemble the second lipid layer, these droplets needed to be transferred through an interface oil-aqueous phase boundary. To achieve this, a high viscosity glycerol solution (50% by volume-in De-ionized water) was placed on top of the diffused oil layer, forming a lipid layer at the interface which gradually sank towards the chip surface (Figure 23: step 3). Upon observing this arrangement after roughly 30 minutes, it was clear that the heavy aqueous media now contained the precipitated aqueous daughter droplets, confirming that a stable lipid bilayer has been formed and the formed bilayer prevents loss of shape and volume of these micro-sized vesicles (Figure 23). This scheme of on-chip formation of large bilayer vesicles is now utilized in conducting a membrane based bio-detection assay.

The two different electrode schemes utilized in this work are shown in Figure 24. Scheme 1 (Figure 24(A(i-iv))) consists of two uniform L-DEP electrode pairs integrated with electrostatic droplet transport electrodes (D-DEP) whereas scheme 2 incorporates continuous tapered and pinched L-DEP electrodes integrated with D-DEP electrodes, to rapidly assemble vesicles and bio-droplets in the size range of 30-150 µm and subsequently manipulate them for bio-assays (Figure 24(B(i-iii))).

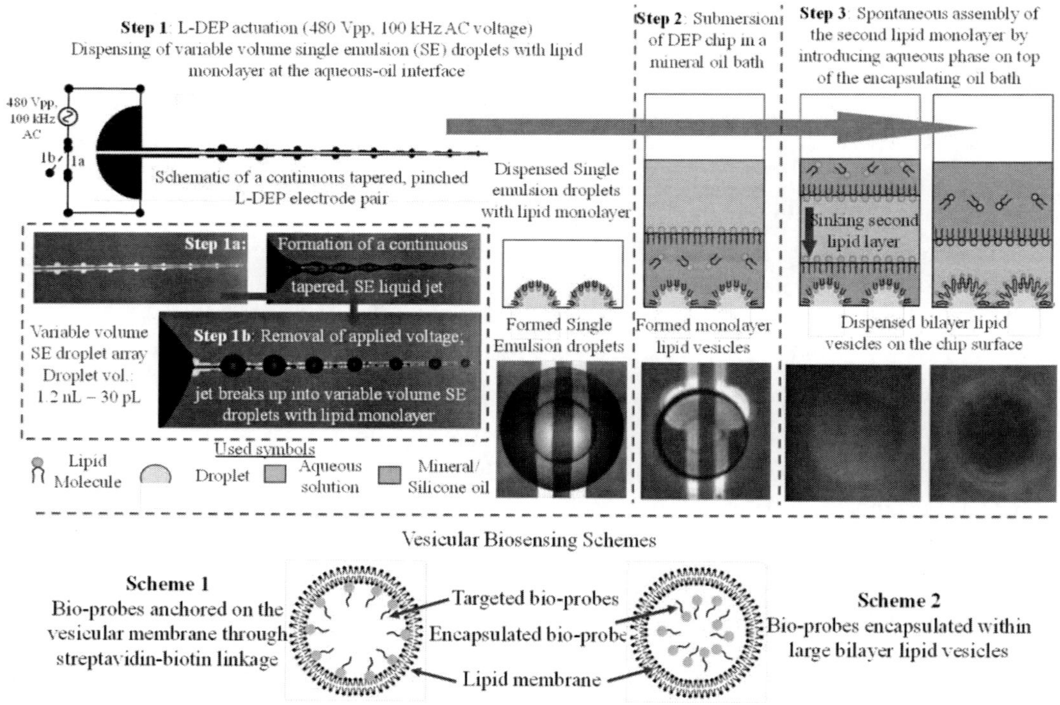

Figure 23. L-DEP SE actuation scheme utilized for dispensing of bilayer lipid vesicles; Schematic of a vesicular bio-detection scheme.

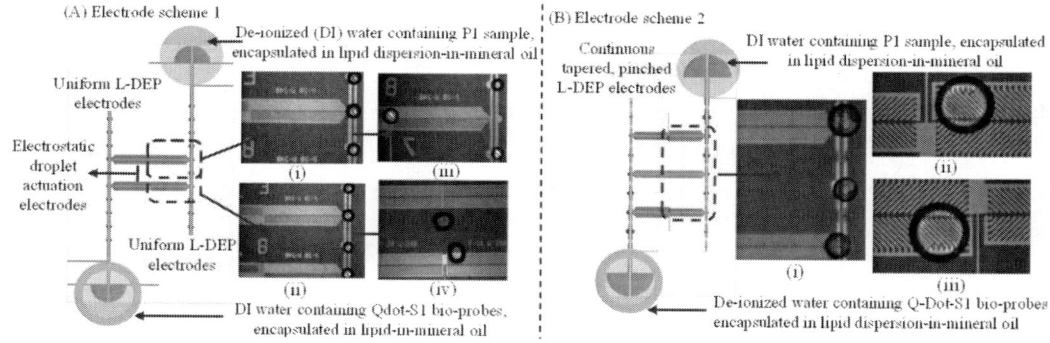

Figure 24. (A) Integrated electrode scheme for fixed size vesicle and bio-sample dispensing and subsequent bio-assays (Scheme 1); A(i, ii) An array of identically sized bio-sample monolayer vesicles after L-DEP actuation; A(iii ,iv) Electrostatic manipulation and mixing of targeted monolayers; (B) Integrated SMF electrode scheme (Scheme 2) for variable size vesicle dispensing and subsequent bio-assays(diameter of the dispensed vesicles: 30-150 µm); B(i) A typical variable sized monolayer vesicle array after L-DEP actuation; B(ii, iii) mixed variable volume bilayer vesicles.

Monolayer and bilayer vesicles were formed using L-DEP actuation of an aqueous sample, encapsulated in the two lipid dispersion media. Vesicles assembled using Lipid 1, when viewed under fluorescent microscope (λ_{ex} = 230 - 400 nm), exhibited a ring shaped green fluorescence (λ_{em} = 525 nm) emanating from the vesicle membrane (Figure 25(a-c)). In contrast, a fluorescent ring was absent when Lipid 2 sample was used to form vesicles. These observations confirm selective anchoring of Qdot® to specific binding sites (biotin heads) in the vesicular membrane.Two bio-assays and a control assay are conducted to demonstrate the applicability, target specificity and detection range of the chip based vesicular bio-sensors. In Assay1, oligonucleotide, S1 was mixed off-chip with Qdot® to prepare a binary mixture containing Qdot® bound oligo bio-probes. Upon L-DEP actuation of this aqueous sample-in-Lipid 1-mineral oil dispersion, the streptavidin linker in the Qdot® molecules allows the Qdot®-S1 bio-probe to anchor itself selectively to the biotin heads in the lipid membrane (Figure 25(d, e)). D-DEP actuation is then used to add a complementary oligonucleotide (P1 with IOWA-FQ™ quencher at 5'-end) daughter droplet, to the functionalized monolayer (Figure 25(a, c, e)).

The mixing results in hybridization of complementary base pairs in the bio-probes and dominant quenching of anchored Qdots (Figure 25(f, g)), measured using PMT (Quenching efficiency (QE) ~ 85%).Two additional assays were also conducted to verify the selectivity and specificity of the binding between the marker and the Qdot® bio-probe. In Assay 2, Lipid 2-in-mineral oil dispersion was used and the Qdot®-oligo bio-probes did not anchor on the membrane (Figure 25(h, i), resulting in quenching of bio-probes only within the vesicle (Figure 25(j, k)), measured using PMT (Quenching efficiency (QE) ~ 75%).

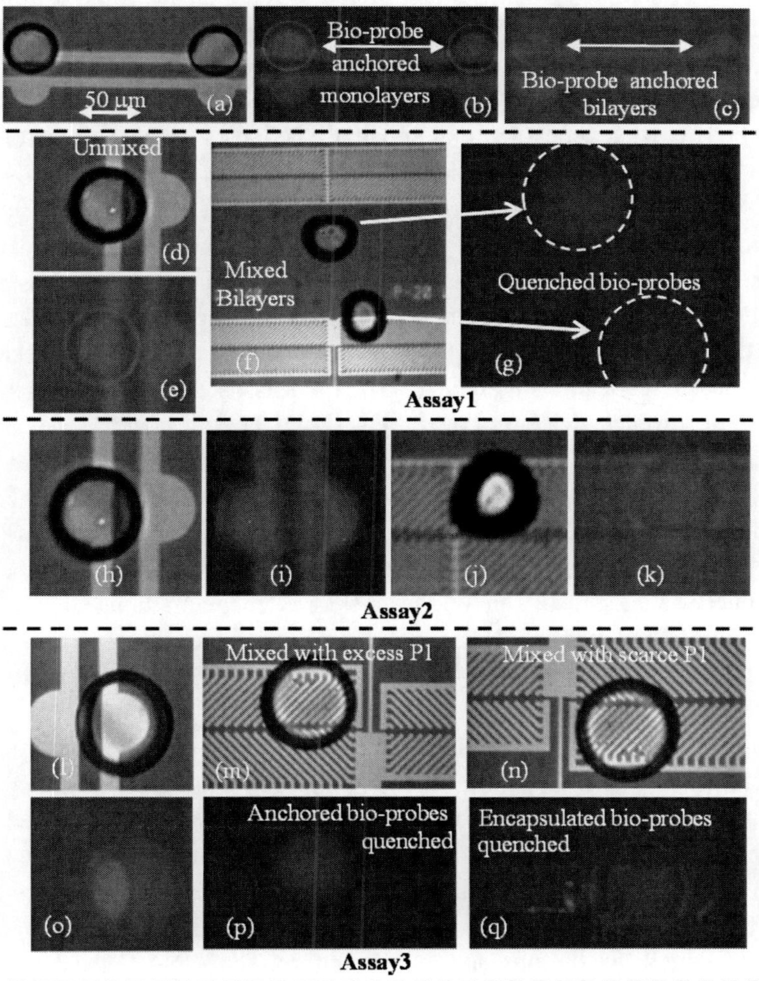

Figure 25. Results of the vesicular bio-detection assays.

In Assay 3, electrode scheme 2 was utilized, where a variable size vesicle dispensing scheme (Figure 24(B)) was used to demonstrate that by controlling the quantity of the oligonucleotide P1 (quencher) in the dispensed variable volume reagent daughter droplets, selective quenching of bio-probes anchored either on the vesicle membrane (with excess amounts of quencher, QE ~ 75%; see Figure 25(m, p)) or, encapsulated within the vesicle (with scarce amounts of quencher, QE ~ 75%; see Figure 25(n, q)) can be achieved. Subsequently, in each assay, a second lipid layer was spontaneously assembled on the monolayer vesicles to secure the bio-probes on/within a more stable bilayer membrane structure (Figure 25(c, e, g), (i, k), (p, q)).

Bead-based bio-assays

In this section, we focus on the capabilities of L-DEP to dispense ultra-low counts of functionalized micro-beads in a controlled fashion, which is then utilized in a bead-based DNA hybridization detection assay [38]. The dynabead® (DB) (diameter: 2.8 μm; Invitrogen, USA) used for this experiment contains an outer shell of streptavidin linkers which readily binds to biotin functional groups. A suitable oligonucleotide sample, S1 with a biotinylated 5'-end and a fluorophore at the 3'-end was bound to the beads to prepare the bio-probe (see Figure 26). Detailed attributes of the binding process, pertaining to preparation of the dynabead® are also reported in [38]. A complementary oligo sample P1 with a quencher molecule attached to its 5' end (see experimental section) was also prepared in identical aqueous media (concentration: 50 μg/mL).

A simple *1x1* electrode scheme, shown in Figure 18(a), is used in the bio-assay. 1 μL parent sample droplets of DB-S1 and P1 are mounted to the two parent sites of the electrode scheme. Upon L-DEP actuation, two arrays of equi-volume daughter droplets are dispensed on the two L-DEP electrodes. Subsequently the D-DEP electrodes are energized and pair of droplets from the two arrays are mixed to facilitate interaction between the bead based bio-probe and its marker sequence (Figure 27(a-d)). PMT is used to measure the fluorescence from the DBs in the droplets dispensed over the L-DEP electrodes $(Ip_{unmixed} - Ip_{bck})$ and in the mixed droplet $(Ip_{mixed} - Ip_{bck})$.

The measured photocurrent data is shown in the caption of Figure 27. This PMT data was used to calculate the associated quenching efficiency (QE) which quantifies the DNA hybridization. Upon successful hybridization, the fluorophore attached to the micro-bead is dominantly quenched which results in a reduced fluorescence emission (QE ~ 90%) from the micro-beads (Figure 27(h, l)). The difference in the fluorescence intensity (Figure 27(f, h) and Figure 27(j, l)) is quantified $(Ip_{unmixed} - Ip_{mixed})$ in order to measure the presence of the complementary DNA molecules.

The micro-bead concentration was lowered (Figure 27(e-h)) during these experiments in order to measure the detection threshold which is limited by the signal loss due to the off-chip detection set-up and the signal-to-noise (Ip_{bck}) ratio. Since the bio-samples used in this assay are fairly simple; assay and detection times were less than a minute. The limit of detection (LOD) for this bead based DNA hybridization assay has been found to be close to ~ 100-200 attomoles for the ultra-low bead assay where a minimum of 2-3 functionalized beads were investigated with the PMT before and after mixing. The DNA hybridization was also quantified in case of high bead concentration and the amount of hybridized DNA probe was found to be ~ 5-10 femtomoles.

The obtained LOD is an improvement when compared to the conventional closed channel DNA hybridization assays (LOD ~ few picomoles) [61], but is higher than the LOD achieved through RT-PCR based amplification assays in microfluidic DNA/RNA detection schemes where numerous PCR cycles (> 10 cycles) are often required to achieve ultra-low LOD. However, one restriction for this ultra-low bead based assay is that it still requires a relatively high DNA conc. (femtomolar) for the parent droplet and the low detection limit is achieved as a direct result of conducting the DNA hybridization on very small bead count (2-3 beads) in the dispensed daughter droplets.

The proposed bead based DNA hybridization assay is most suited for post amplification screening assays where off-chip or, on-chip RT-PCR amplicons can be screened with labeled bead based target sequences, while using very small amounts (few microliters) of expensive PCR products and marker sequences. Different probe DNA concentrations were investigated using the bead based assay using the ultra-low bead concentration (2-3 beads/nL droplet), shown in Figure 27(e-j), and the quantitative results are reported in Table 4.

The reduced QE corresponding to lower probe concentration is a result of poor FRET based quenching at relatively low quencher concentrations (Bead-bound S1 concentration ~ 0.5 μg/mL). A control assay was also conducted to validate that the loss of fluorescence intensity and the resulting QE were indeed as a result of detection of hybridization of complementary DNA sequence.

P1: 5'-IOWA FQ™ quencher; 20 bases with 3'-biotinylation

5- /5IAbFQ/TGATGGTGCCTATTTTACGG/3Bio/-3

S1: 5'-biotinylated; 20 bases complementary to P1 for hybridization; 6-FAM as fluorophore

5-/5Biosg/CCGTAAAATAGGCACCATCA/36-FAM/-3

S2: 5'-biotinylated; 20 bases non-complementary to P1 for no hybridization; 6-FAM as fluorophore

5-/5Biosg/CAGTAAAATAGTCTCTGCTT/36-FAM/-3

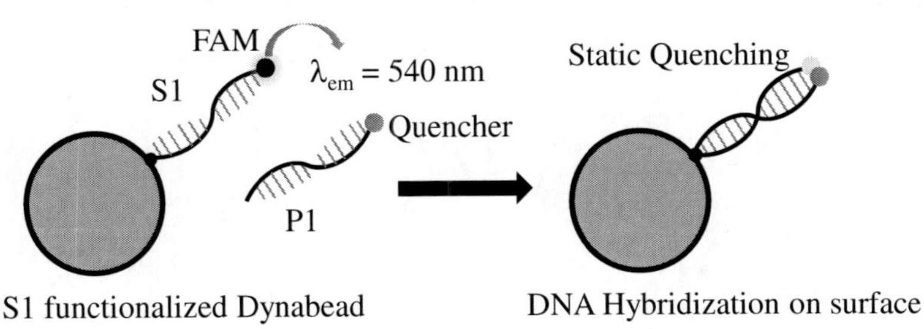

Figure 26. Properties of the used DNA samples and schematic of the bead based DNA hybridization.

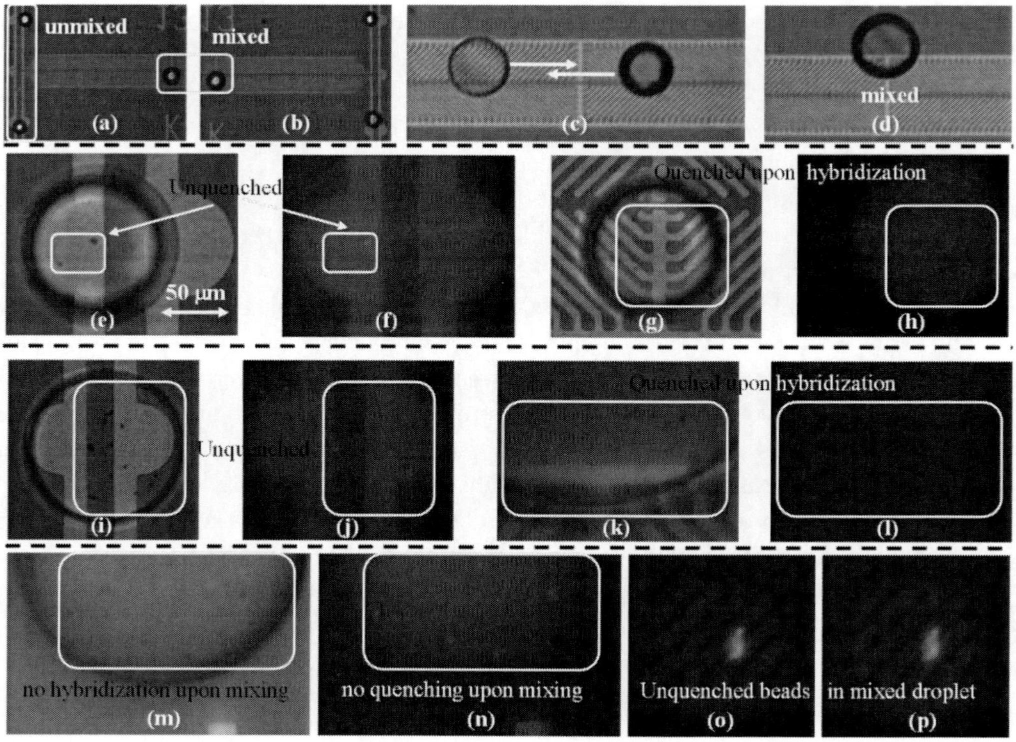

Figure 27. Snapshots showing experimental results from on-chip bead based bio-assays. (a, b) shows mixed and unmixed sample/reagent droplets; (c, d) unmixed daughter droplets with functionalized dynabeads® and corresponding markers prior to and after binary mixing; (e-h) micrographs showing bead based assay achieved at ultra-low bead conc.(2-3 beads per droplet); (e, f) bright and fluorescent image of unmixed and unquenched functionalized micro-beads ($Ip_{unmixed} - Ip_{bck} = 8.6x10^{-6}$ A); (g, h) quenched micro-beads in mixed droplet (QE ~ 91%) as a result of DNA hybridization ($Ip_{mixed} - Ip_{bck} = 0.9x10^{-6}$ A); (i-l) similar bead based assay results for relatively larger bead conc. (~ 10 beads/nL) with ($Ip_{unmixed} - Ip_{bck} = 2.2x10^{-5}$ A), ($Ip_{mixed} - Ip_{bck} = 2.6x10^{-6}$ A) and QE ~ 88%; (m-p) results of a control experiment showing mixing but no quenching for non-specific bio-marker at low bead concentration and 50X optical magnification ($Ip_{mixed} - Ip_{bck} = 6.8x10^{-6}$ A) and QE ~ 20-25% (m, n) and high bead concentration (o, p) ($Ip_{mixed} - Ip_{bck} = 7.0x10^{-6}$ A) and QE ~ 20-25%.

Table 4. Quantitative analysis of probe (P1) DNA hybridization with ultra-low bead aliquots

P1 conc. (μg/mL)	$Ip_{unmixed}$ (μA)	Ip_{mixed} (μA)	Ip_{bck} (μA)	QE (%)	Hybridized P1(amol)
0.5	7.55	3.97	1.04	55%	120
1	7.50	3.29	1.05	65%	180
10	7.60	2.04	1.06	85%	180
25	7.52	1.8	1.05	88%	120
50	7.55	1.7	1.05	91%	180

In the control experiment, a non-complementary marker oligo sample S2 (see Figure 27(m-p)) was used to functionalize the DB micro-beads.

The assay was repeated and the mixed and unmixed droplets were analyzed using the PMT. The resultant mixed droplets for this assay, for both low and high bead conc. (shown in Figure 27(m-p)) suggest very little change in fluorescent intensity (QE < 20-25%) of the bead based bio-probes which confirms that the particular sequence of marker was not present in the sample oligonucleotide.

CONCLUSION

In this chapter, we demonstrated the utility of DEP based miniaturized liquid sample handling and its illustrative lab on chip applications. The underlying principles of various DEP based liquid and particle actuation schemes were elaborated upon in the theoretical section of the chapter. Suitable device fabrication protocols for reliable device design are also detailed in the fabrication section. L-DEP based SMF schemes were implemented to achieve droplet dispensing, transport and manipulation in the volume regime of nL to pL. L-DEP was also utilized in dispensing artificial vesicular membranes and controlled distribution of functionalized micro-beads.

Applications such as DNA quantification assay, DNA hybridization assay and Bead-based DNA assay were demonstrated using various integrated DEP electrode schemes. Hence, L-DEP was established as a pioneer surface microfluidic technology for various LOC based bio-applications.

FUTURE WORK

A number of issues need to be addressed in order to realize practical DEP based LOC devices that can be applied and can handle, process and analyze the more complex real clinical samples such as blood, saliva and other biofluids.

For example, a self-contained LOC device capable of genetic analysis requires a large scale integration (LSI) of various droplet processing (Figure 28), detection and analysis blocks. This would include nucleic acid amplification, requiring mixing of sample and reagents followed controlled thermal cycling of the sample droplets.

To detect on-chip the presence of the amplified target nucleic sequence would require the hybrid integration of a substrate housing the optical sources, low-loss waveguides, filters, sensitive near field wideband optical detectors and high and low voltage electronics to control chip operation with the surface microfluidic manipulation chip platform.

The reliable manipulation of biological samples containing nucleic acids, proteins and enzymes requires that the surfaces be suitably modified to minimize adsorption while at the same time maintaining high degree of hydrophobicity.

The control of the surface wettability in such cases poses a considerable challenge, which may be overcome by creating a super hydrophobic surface by Nano-texturing the surface. The DEP actuation of the more conductive bio-samples requires the use of much higher frequency fields and thus very high slew rate and high voltage amplifier to drive the DEP actuation electrodes. Hence in such cases the performance of the amplifier will have a direct impact on the ability to manipulate such samples. Thus a number of challenges remain that need to be addressed in the future in order to realize and deploy DEP based LOC devices, capable of handheld analysis of biosamples.

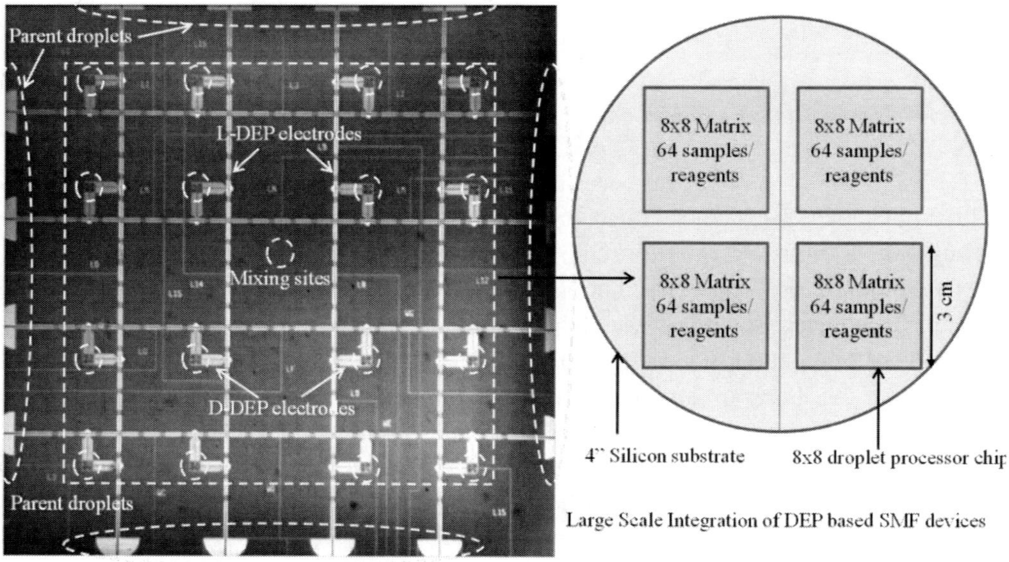

Figure 28. Four LSI DEP based SMF devices fabricated on a single 4” Silicon wafer.

ACKNOWLEDGMENTS

The authors acknowledge the financial support from NSERC CANADA and CMC Microsystems CANADA for the presented research work. The microfluidic chips used in the reported work were fabricated at Nanofab, University of Alberta, CANADA and the experimental and numerical analysis was conducted at the BioMEMS Lab, University of Calgary. The authors also acknowledge useful discussions and feedback provided by Dr. Thomas B. Jones (University of Rochester), Dr. Masao Washizu (University of Tokyo) and Dr. Reginald Paul (University of Calgary) during the course of this research work. The authors furthermore recognize the valuable contribution of Dipanker Chugh and Thiru Thirukumaran in the development of the surface microfluidic device technology.

REFERENCES

[1] Manz, A., Graber, N., and Widmer, H. M. (1990). Miniaturized total chemical analysis systems: A novel concept for chemical sensing. *Sensors and Actuators-B,* vol. 1, pp. 244-248.

[2] Whitesides, G. M. (2006). The origins and the future of microfluidics. *Nature,* vol. 442, pp. 368-373.

[3] Duffy, D. C., McDonald, J. C., Schueller, O. J., and Whitesides, G. M. (1998). Rapid prototyping of microfluidic systems in poly(dimethylsiloxane). *Analytical Chemistry,* vol. 70, pp. 4974-4984.

[4] McDonald, J. C., and Whitesides, G. M. (2002). Poly(dimethylsiloxane) as a material for fabricating microfluidic devices. *Accounts of Chemical Research,* vol. 35, pp. 491-499.

[5] Unger, M. A. et. al. (2000). Monolithic microfabricated valves and pumps by multilayer soft lithography. *Science,* vol. 288, pp. 113.

[6] Thorson, T., Maerkl, S. J., and Quake, S. R. (2002). Microfluidic large-scale integration. *Science,* vol. 298, pp. 580.

[7] Fenniri, H., and Alvarez-Pubela, R. (2007). High-throughput screening flows along. *Nature Chemical Biology,* vol. 3, pp. 247-249.

[8] Thorsen, T. A. (2004). Microfluidic tools for high-throughput screening. *J. Biotechniques,* vol. 36, pp. 197-199.

[9] Hong, J. W., and Quake, S. R. (2003). Integrated nanoliter systems. *Nat. Biotechnology,* vol. 21, pp. 1179–1183.

[10] Madou, M., Zoval, J., Jia, G., Kido, H., Kim, J., and Kim, N. (2006). Lab on a CD. Annual Review of *Bio*medical Engineering, Vol. 8, pp. 601-628.

[11] Chen, et. al. (2007). A wettability switchable surface by microscale surface morphology change. *J. Micromech. Microeng,* vol. 17, pp. 489–495.

[12] Darhuber, A. A., et. al. (2003). Thermocapillary actuation of droplets on chemically patterned surfaces by programmable microheater array. *J. MEMS,* vol. 12, pp. 873-879.

[13] Wixforth, A. (2006). Acoustically Driven Programmable Microfluidics for Biological and Chemical Applications. *JALA,* vol. 11(6), pp. 399-405.

[14] Langelier, et. al. (2009). Acoustically driven programmable liquid motion using resonance cavities. *proc. National Academy of Sciences of USA,* vol. 106(31), pp. 12617-12622.

[15] Lippmann, M. G. (1875). Relation entre les phénomènes électriques et capillaries. *Ann. Chim. Phys,,* vol. 5: pp. 494.

[16] Moon, H., et. al. (2002). Low voltage electrowetting on dielectric. *J. Applied Physics,,* vol. 92, pp. 4080.

[17] Mugele, F., and Baret, J. C. (2005). Electrowetting: from basics to applications. *J. Phys.: Condens. Matter,,* vol. 17, pp. R705–R774.

[18] Pollack, M. G., Shenderov, A. D., and Fair, R. B. (2002). Electrowetting-based actuation of droplets for integrated microfluidics. *Lab Chip.,* vol. 7152, pp. 96–101.

[19] Pohl, H. A. (1958). Some effects of non-uniform fields on dielectrics. *J. Appl. Phys.,*vol. 29(8), pp. 1182 – 1188.

[20] Pellat, H. (1895). Mesure de la force agissant sur les dielectriques liquids non electrises places dans un champ elitrique. *C R Acad Sci, Paris.,* vol. 119, pp. 691–694.

[21] Jones, T. B., Perry, M. P., and Melcher, J. M. (1971). Dielectric siphon. *Science.,* Vol. 174(4015), pp. 1232 – 1233.

[22] Jones, T. B. (2001). Liquid dielectrophoresis on the microscale. *J. Electrostatics.,* vol. 51– 52, pp. 290–299.

[23] Davis, J. M., and Giddings, J. C. (1986). Feasibility study of dielectrical field-flow fractionation. *Sepa. Sci. and Tech.,* vol. 21, pp. 969-989.

[24] Gascoyne P. R. C. et. al., "Dielectrophoretic separation of mammalian cells studied by computerized image analysis". *Meas. Sci.Technol.* vol. 3, pp. 439-445, 1992.

[25] Wang, X. B., et.al. (1998). Separation of polystyrene microbeads using dielectrophoretic/gravitational field-flow-fractionation. *J. Biophys.,* vol. 74, pp. 2689–2701.

[26] Hosseini Y., Hartley L. F. and Kaler K. V. I. S., "Hybrid Integrated CMOSMicrofluidic Device for the Detection and Characterization of Particles", Procc. IEEE MNRC, 2008.

[27] Jones, T. B., Gunji, M., Washizu, M., and Feldman, M. J. (2001). Dielectrophoretic liquid actuation and nanoliter droplet formation. *J. Appl. Phys.,* vol. 89(3), pp. 1–8.

[28] Ahmed, R., and Jones, T. B. (2006). Dispensing picoliter droplets on substrates using dielectrophoresis. *J. Electrostatics.,* vol. 64, pp. 543–549.

[29] Kaler, K. V. I. S., Prakash, R., and Chugh, D. (2010). Liquid dielectrophoresis and surface microfluidics. *Biomicrofluidics,*4(2), pp. 1-17(022805).

[30] Gunji, M., Nakanishi, H., and Washizu, M. (2004). Droplet actuation based on single-phase electrostatic excitation. *Proc. of Micro Total Analysis Systems,* 2004, vol. 1, pp 168–170.

[31] Gunji, M., and Washizu M. (2005). Self propulsion of a water droplet in an electric field. *J. Phys. D: Applied Physics,* vol. 38(14), pp. 2417.

[32] Jones, T. B. (2001). Dynamics of Dielectrophoretic Liquid Microactuation. proc. 4th Int'l Conference on Applied Electrostatics, Dalian, China.

[33] Thirukumaran, T. K., and Kaler, K. V. I. S. (2007). Surface microfluidics—high-speed DEP liquid actuation on planar substrates and critical factors in reliable actuation. *J. Micromech Microeng.,* vol. 17, pp. 743–752.

[34] Chugh, D., and Kaler, K. V. I. S. (2010). Integrated liquid and droplet dielectrophoresis for biochemical assays. *Microfluid Nanofluid,* vol. 8, pp. 445-456.

[35] Prakash, R., and Kaler, K. V. I. S. (2009). DEP actuation of emulsion jets and dispensing of sub-nanoliter emulsion droplets. *Lab Chip.,* vol. 9(19), pp. 2836–2844.

[36] Prakash R., Paul R., and Kaler, K. V. I. S. (2010). Liquid DEP Actuation and Precision Dispensing of Variable Volume Droplets. *Lab Chip,* 10, pp. 3094-3102.

[37] Prakash R., and Kaler, K. V. I. S. (2010). Chip Based Unilamellar Vesicle Formation and Dispensing Using Dielectrophoresis. *Proc. MicroTAS,* vol. 1, pp. 417-419.

[38] Prakash R., and Kaler K. V. I. S. (2012). Microbead dispensing and ultralow DNA hybridization detection using liquid dielectrophoresis. *Sensors and Actuators B: Chemical,* vol. 169, pp. 274-283.

[39] Lord Rayleigh (1879). On capillary phenomena of jets. *Proc. Roy. Soc.* (London), vol. 27, pp. 71–97.

[40] King, M. R. Lomakin, O. A., Ahmed, R., and Jones, T. B. (2005). Size-selective deposition of particles combining liquid and particulate dielectrophoresis. *J. App. Phys.,*vol. 97(054902), pp. 1-7.

[41] White, C. M., Holland, L.A., and Famouri, P. (2010). Application of capillary electrophoresis to predict crossover frequency of polystyrene particles in dielectrophoresis. *Electrophoresis,* vol. 31, pp. 2664–2671.

[42] Park, S., et. al. (2011). Continuous dielectrophoretic bacterial separation and concentration from physiological media of high conductivity. *Lab chip.,* vol. 11, pp. 2893-2900.

[43] Vink, et. al. (1995). On the homogeneity of sputter-deposited ITO films Part I. Stress and microstructure. *Thin Solid Films,* vol. 266, pp. 145-151.

[44] Synowicki, R. A. (1998). Spectroscopic ellipsometry characterization of indium tin oxide film microstructure and optical constants. *Thin Solid Films,* vol. 313-314, pp. 394–397.

[45] Prakash, R., Kaler, K. V. I. S., Papageorgiou, D. P., and Papathanasiou A. G. (2012) Performance of Multilayered Fluoropolymer Surface Coating for DEP Surface Microfluidics Devices. *Microfluid. Nanofluid., DOI*: 10.1007/s10404-012-0963-1.

[46] Papageorgiou, D. P., Koumoulos E. P., Charitidis C. A., Boudouvis A. G. and Papathanasiou, A. G. (2011). Evaluating the Robustness of Top Coatings Comprising Plasma-Deposited Fluorocarbons in Electrowetting Systems. *J. Adhes Sci Technol.,* doi:10.1163/ 156856111X600226.

[47] Prakash, R., Chugh, D., and Kaler, K. V. I. S. (2010). Quantitative Biochemical Assay On A Surface Microfluidic Device. *Proc. 3rd International Joint Conference on Biomedical Engineering Systems and Technologies* (Biodevices 2010), pp. 51-58.

[48] Burns, M. A. et. al. (1998). An integrated nanoliter DNA analysis device. *Science vol.,* 282(5388), pp. 484–487.

[49] Liu, R. H., et. al. (2004). Self-contained, fully integrated biochip for sample preparation, polymerase chain reaction amplification, and DNA microarray detection. *J. Anal. Chem., vol.,* 767), pp. 1824–1831.

[50] Mathies, R. A., and Lagally, E. T. (2004). Integrated genetic analysis microsystems, *J. Phys. D: Appl. Phys. Vol.,* 37, pp. R245–R261.

[51] Heller, M. J. (2002). DNA Microarray technology: Devices, systems and applications. *Annu. Rev. Biomed. Eng.,* vol. 4, pp. 129.

[52] Singer, V. L., Laurie, J. J., Yue, S. T., and Haugland, R. P. (1997). Characterization of PicoGreen reagent and development of a fluorescence-based solution assay for double-stranded DNA quantitation. *Anal Biochem,,* vol. 249, pp. 228–238.

[53] Didenko, V. V. (2001). DNA probes using fluorescence resonance energy transfer (FRET): designs and applications. *Biotechniques,* vol. 31, pp.1106–1121.

[54] Prakash, R., Chugh, D., and Kaler, K. V. I. S. (2009). Quantitative DNA Hybridization Assay On A Multiplexed Surface Microfluidic Device. *Proc. 2nd Micro/Nanoscale Heat Transfer International Conference (*MNHMT 2009), pp. 1-9.

[55] Massey, M., Algar, W. R., and Krull, U. J. (2006). Fluorescence resonance energy transfer (FRET) for DNA biosensors: FRET pairs and Förster distances for various dye–DNA conjugates. *Analytica Chemica Acta,* vol. 568, pp. 181-189.

[56] Marras, S. A. E., Kramer, F. R., and Tyagi, S. (2002). Efficiencies of Fuorescence resonance energy transfer and contact-mediated quenching in oligonucleotide probes. *Nucleic Acids Research,* vol. 30(21), pp. e122.

[57] Pautot, S., Frisken, B. J., and Weitz, D. A. (2003). Engineering Asymmetrical Vesicles. *PNAS,* vol. 100(19), pp. 10718-10721.

[58] Stachowiak, et. al. (2008). Unilamellar vesicle formation and encapsulation by microfluidic jetting. *PNAS,* vol. 108(12), pp, 4697-4702.

[59] Hadorn, M., and Hotz, P. E. (2009). Multivesicular Assemblies as Real-World Testbeds for Embryogenic Evolutionary Systems. Proc. the 4th Australian Conference on Artificial Life: Borrowing from Biology, Melbourne, Australia, pp. 169-178.

[60] Prakash, R., and Kaler, K. V. I. S. (2011). Chip Based Assembly of Vesicular Biosensors Using Quantum Dots as Bio-probes. *Proc. MicroTAS* (2011), vol. 1, pp. 1448-1450.

[61] Krishnan J., Kim, T. S., and Kim, S. K. (2008). Microfluidic Superparamagnetic Bead-Based Multiplex Detection System. *Proc. MicroTAS* 2008, pp. 1462-1464.

In: Microfluidics
Editors: Stefano Panzarella and Walter Maroni
ISBN: 978-1-62417-345-5

Chapter 2

MICROFLUIDIC BIOSENSOR SYSTEMS FOR CELL BIOLOGY AND DRUG DISCOVERY

Ye Fang[*]
Biochemical Technologies, Science and Technolog
Division, Corning Inc., Corning, New York, US

ABSTRACT

The ability to translate drug-target interactions into cell phenotypic signatures has made label-free biosensors become a powerful platform for the molecular delineation of receptor signaling and drug pharmacology in native cells. Microfluidics enables miniaturization of assays, increases experimental throughput, and allows precise control of cell phenotypes, molecular concentrations and drug-target interactions. Integration of microfluidics with label-free biosensors would offer additional dimensions for investigating the cellular and molecular mechanisms of receptor signaling and drug actions down to single cell level. This chapter first reviews key aspects of microfluidics for studying cell biology, and discusses the recent advances and applications in cell biology and drug discovery of microfluidic biosensor devices.

Keywords: Cell signaling; electric biosensor; G protein-coupled receptor; label-free biosensor; microfluidics; optical biosensor; resonant waveguide grating biosensor; single cell analysis

INTRODUCTION

The past decades have witnessed the increasing use of cell-based assays to examine drug-target interactions and their functional consequences in living cells, which is coincident with

[*] Corresponding author: Ye Fang. Biochemical Technologies, Science and Technology Division, Corning Inc., Corning, New York 14831, United States of America. E-mail: fangy2@corning.com.

the expansion of druggable targets and the improved resolution and throughput of detection technologies.

Our ever-improving understanding of the molecular basis of human diseases, together with advances in molecular genetics, have led to identification of many novel druggable targets that are critical to regulate the function and behaviors of cells and cell systems [1, 2]. Advances in molecular biology have made it possible to manipulate many distinct cellular targets, one at a time so the biological functions of these targets can be manifested [3-7]. Advances in detection and instrumentation have made it reality to directly visualize the spatial and temporal dynamics of a wide variety of cellular processes at the molecular resolution in living cells [8, 9], as well as to assay thousands of biological reactions within a single experiment [10, 11].

Cell-based assays can be classified into molecular and phenotypic assays [12]. Molecular assays rely on measurements of specific molecules in the receptor signaling cascades to infer the molecular mode of action of drug molecules [13]. Molecular assays are advantageous in terms of screening and medicinal chemistry optimization, but are limited for predicting *in vivo* impacts of drugs [14].

In contrast, phenotypic assays measure a specific physiological or cellular phenotype in physiologically relevant cells or cell systems to determine the activity of drug molecules [12]. These phenotypic assays may be able to directly measure a pharmacological activity of a drug linking to its therapeutic impact for a given disease state [15]. However, in many cases the physiological or cellular phenotypes measured are too complex to derive mechanistic descriptions of drug actions, thus making the phenotypic assays impracticable to medicinal chemistry optimization [16].

Historically, drug discovery has been heavily weighted toward phenotypic assays using isolated animal tissues to determine drug properties without detailed understanding of the physiological origin of the pharmacologic response measured [16].Since genomics and other omics emerged two decades ago, phenotypic assays have given in for molecular assays, later of which have become dominating in both basic research and drug discovery due to considerations associated with operational and biological simplicity. With mixed success of target-centric, molecular assay-enable drug discovery [12] has led to renewed interests in phenotypic assays with mechanistic deconvolution capability.

In recent years, cell-based assays enabled by label-free biosensors have reemerged as novel phenotypic pharmacologic assays for cell biology and drug discovery, owing to the ability of these biosensors to translate drug-target interactions in living cells into whole cell phenotypic signatures as well as to deconvolute molecular and cellular mechanisms of receptor signaling and drug activity [13, 16].

In parallel, microfluidics has gained increasing acceptance in the field of cell biology and drug discovery in the past decade [17-20]. These microsystems enable accurate flow control in the spatial and temporal domains [21, 22], create new revenue for precise control of cell growth and manipulation [23-25], and allow integration with a wide variety of detection technologies for on-chip quantitative analysis of the output of processes and molecular interactions within a single cell [26, 27].

Thus, integration of microfluidics with label-free biosensors would offer additional dimensions for investigating the cellular and molecular mechanisms of receptor signaling and drug activity. This chapter reviews recent developments in label-free biosensor whole cell assays and microfluidic biosensor systems for cell biology and drug discovery.

LABEL-FREE BIOSENSORS FOR CELL ANALYSIS

A biosensor is a device consisting of three critical components, a biological element, a detector element, and a transducer associated with both elements. The biosensor uses a transducer to convert a molecular recognition event or a stimulus-induced cellular response into a quantifiable signal (i.e., biosensor signal). Owing to distinct operational principles, different biosensors have been used to study different physical properties of cells (Figure 1).

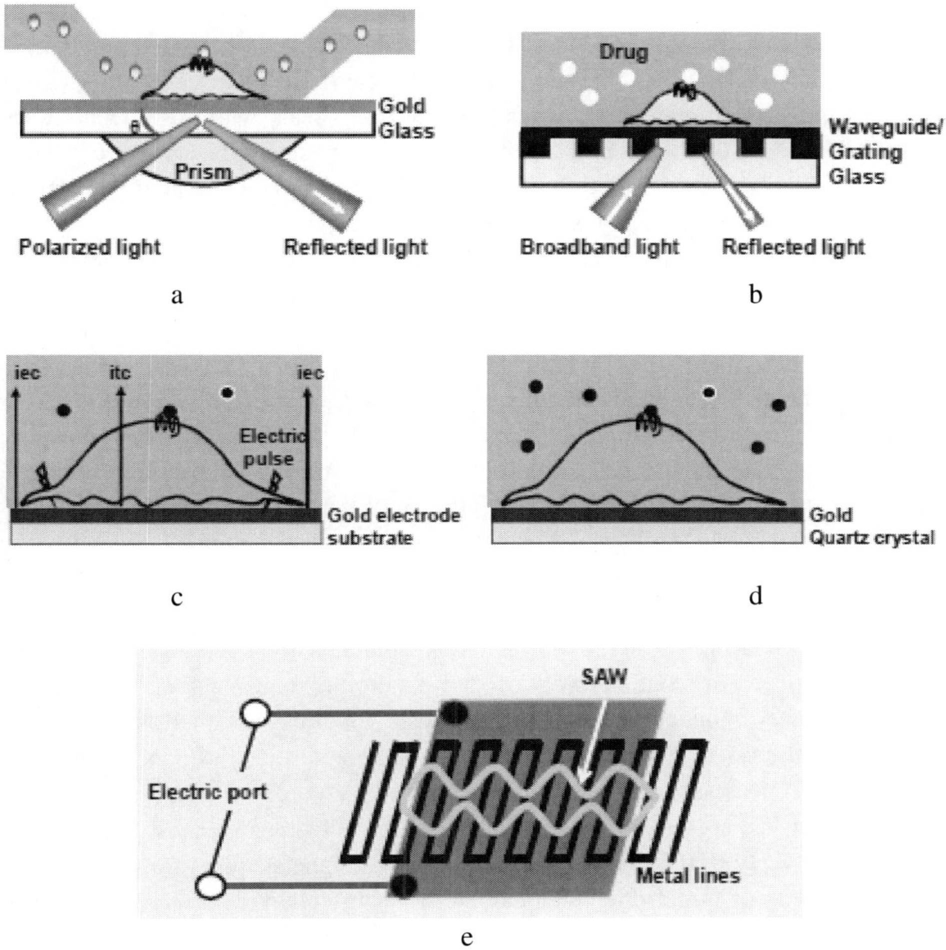

Figure 1. Principles of label-free biosensors for whole cell sensing. (a) Surface plasmon resonance (SPR) which uses a microfluidics device to deliver ligands to the cells immobilized on the surface of gold thin film, so ligand-induced dynamic mass redistribution (DMR) in living cells can be monitored in real time. (b) Resonant waveguide grating (RWG) which uses a nanograting to couple light into the waveguide via diffraction, so ligand-induced DMR in living cells can be monitored in real time. (c) Electric biosensor which uses a low electrolyte impedance interface to detect both flows of extracellular (iec) and transcellular (itc) currents, so the ionic environment surrounding the biosensor and the cells can be monitored in real time. (d) Quartz crystal microbalance (QCM) which uses an oscillating crystal to detect ligand-induced changes in cellular mass and viscoelastic property. (e) Surface acoustic wave (SAW) which uses a surface acoustic wave propagating in the lateral direction on a chip made from a mono-crystalline piezo-electric material to detect ligand-induced changes in cellular mass and viscoelastic property.

Optical biosensors including surface plasmon resonance (SPR) and resonant waveguide grating (RWG) use a surface bound evanescent wave generated via different mechanisms to characterize alterations in local refractive index at the sensor surface [28]. Electric biosensors use a low electrolyte impedance interface to detect the changes in the impedance of the biosensor system under electric fields generated with sinusoidal voltages [29].

Atomic force microscopy uses a mechanical cantilever to sense the changes in the biomechanics of cells [30, 31]. Coherent anti-stokes Raman spectroscopy (CARS) and stimulated Raman scattering (SRS) microscopy employ Raman scattering to examine the production and organization of molecules having specific vibrational signatures in cells [32, 33]. SRS uses two laser beams tuned to different frequencies, so if the difference in frequency between the beams matches the vibration of specific chemical bonds, the Stokes beam will experience stimulated Raman gain, whereas the pump beam will experience stimulated Raman loss. The measurements of Raman scattering gain and loss can be used to image specific molecules in tissues and cells. Light microscopy directly images the density and organization of biomolecular complexes in cells [34].

Quartz crystal microbalance (QCM) reads the resonant frequency and resistance of a quartz crystal to determine the mechanical and hydrodynamic properties of cells [35]. Surface acoustic wave (SAW) uses an acoustic wave traveling along the surface of a material exhibiting elasticity to manipulate and sense cells [28, 36, 37].

The origin of biosensor signals arising from living cells upon stimulation is divergent and dependent on the detection principles of biosensors. The optical signal obtained using RWG is a function of dynamic mass redistribution (DMR) within the cells [13]. The impedance signal obtained using electric biosensors is a function of the morphological changes and ionic redistribution of cells [13, 29]. The AFM signal is a function of the interaction between probe molecules attached to the cantilever tip and a specific cell surface molecule, and the motion and elastic properties of living cells [30]. The vibrational signatures and images obtained using CARS and SRS are mostly from relatively abundant molecules such as unsaturated fatty molecules in cells or from naturally occurring chromophores such as haemoglobin [32, 33]. The resonance frequency-resonance resistance diagram obtained using QCM is a function of mass and viscoelastic property on the biosensor surface [38]; and the similar is for the acoustic wave phase-shift signal obtained using SAW [28]. Except for Raman-based microscopy which has specificity to certain molecules, all other biosensors measure a cell phenotypic response. Both RWG and SPR monitor DMR signals arising from the downstream cell signaling events of target-drug interactions in living cells [13, 39-44]. Several cellular and molecular deconvolution studies suggest that a drug-induced DMR consists of contributions from several cellular events downstream the drug-target interaction, including protein trafficking, actin remodeling, alteration in cell adhesion and eventually cell morphological changes [41-45].

Different biosensors are also divergent in their spatial and temporal resolution. Optical, electric, QCM and SAW biosensors all allow real-time kinetic measurements of cellular responses with relatively high temporal resolution. The real-time observation of drug effect on native cells offers advantages in that multiplex signaling pathways can be decoded through the use of multiple kinetic parameters obtained [40]. However, these biosensors generally measure an averaged response from a population of cells [13]. The newer SPR and RWG imaging systems have relatively high spatial resolutions that can resolve single cells [46-48]. CARS and SRS microscopies have slightly higher resolution than light microscopy, so they can resolve specific organelles [32, 33]. AFM has high resolution for detecting cell surface features, and in the best case, can visualize the structural features underneath the cell surface [30, 49].

Different biosensors have different pathway coverage. Optical, electric and QCM biosensors are believed to have broad pathway coverage. In recent years, advances in biosensor design and instrumentation have expanded the applications of both RWG and electric biosensors to examine a diverse array of cellular processes, including cell adhesion, cell barrier functions, proliferation and toxicity, cell-cell communication, cell signaling, cell differentiation, cell migration, muscle cell hypertrophy, cardiomyocyte beating, and viral infection [reviewed in 50]. Furthermore, DMR assay enabled by RWG biosensors has been shown to have high sensitivity and wide pathway coverage, so it is possible to investigate a wide range of endogenous receptors in native cells including stem cells and primary cells [51]. The biosensor signals obtained using these biosensors are integrative in nature, thus providing a holistic view of receptor signaling and drug pharmacology [52]. Furthermore, these non-invasive biosensors allow cellular and molecular delineation of receptor signaling when combining with gene manipulation techniques and probe molecules [53].

MICROFLUIDIC BIOSENSOR SYSTEMS

Microfluidic devices have been developed for advanced cell culture and cell analysis. These devices can be fabricated using lithographic technologies on glass or silicon surfaces, or replica molding of cell culture compatible polymers from patterned template tools, or injection molding of cell culture compatible polymers. As a transparent, affordable and biocompatible elastomer, poly(dimethylsiloxane) (PDMS) is the most widely used material for soft lithographic molding of microfluidic device [54]. PDMS is unique in that it has excellent gas permeability, and is easy and affordable to form diverse patterns by soft lithography and amenable to chemical modification [55]. However, PDMS does have limitations associated with its innate hydrophobicity, high retention for certain drugs and absorption of organic solvents, and evaporation of water [56]. To overcome this, alternative materials including polymethylacrylate, cyclic olefin polymers, thermoset polyester, elastomers, photocurable hydrogels, and photocurable solvent-resistant elastomers such as perfluoropolyethers have been also explored [26, 56]. Biodegradable elastomers including poly(dl-lactide-co-glycolide) (PLGA) and poly(glycerol sebacate) (PGS) are also useful to microfabricate capillary networks [57]. Owing to its rigid mechanical properties and bulk degradation kinetics, PLGA-based microfluidic devices are mostly useful for short term cell culture [58].

Realizing the full potential of microfluidics-centric lab-on-a-chip devices requires integration and packaging of several functionalities into a single system. Optofluidic technologies combining microfluidics with optics have gained momentum in cell research, drug discovery and medical diagnostics. However, most optic techniques used for cell research require labels and/or genetic manipulations. Label-free detection including optical and electric biosensors represents a potentially powerful alternative to fluorescent and luminescent detection. The integration of microfluidics and label-free biosensors should be able to gather precise biochemical and mechanistic data from cells and tissues.

SPR is well-known for kinetic analysis of biomolecular interactions. SPR generally uses a microfluidics device to control the association and dissociation of drug molecules binding to the target immobilized on the surface of a gold thin film [28].

RWG is an alternative optical biosensor for biomolecular interaction analysis. In an effort to increase throughput for binding analysis, Choi and Cunningham [59] used a nanoreplica molding process to produce a 96well RWG biosensor microplate device having polymeric microfluidic channels. In this device, a central ''common'' well in each row is used as an access point for introduction or withdrawal of reagents for the flow channels. The flow channel has serpentine patterns, enabling an equal length flow path from each analyte well to the common well. Pneumatic fluid driving force is used to deliver or remove solutions to or from the sample wells, respectively. Scanning wavelength interrogation detection system is used to detect biomolecular interactions occurring at each sensor along the scanning path in a sequential manner. The microplate supports, in theory, analysis of the interactions of up to eight distinct ligands with eleven different receptors, or one ligand with eighty-eight different receptors. However, it appears that the system has insufficient temporal resolution to accurately determine the on and off rates of biomolecular interactions. The applications of this device for cell-based assays have not been demonstrated.

Recently, we had developed a RWG imager that is based on swept wavelength interrogation scheme, wherein the light beam from a swept tunable light source is directed to illuminate the biosensor array, and the reflected light is recorded using a high speed CMOS (complementary metal-oxide semiconductor) camera (Figure 2a). In this system, total 150 spectral images are acquired every 3 sec during a single cycle of wavelength sweep from 825 to 840 nm, and the spectral image stack is then processed into sensor resonant wavelength (Figure 2b), or DMR image (Figure 2c). The imager can image simultaneously all biosensors within the field of view and has a high temporal resolution of 3 sec for kinetic measurements. The imager has different spatial resolutions, normally in the range of 12 to 80 μm, depending on the size of the field of view [47, 48].

We also described two different microfluidic RWG biosensor systems, both of which consist of three key components: an array of functional RWG biosensors, a microfluidic device, and a RWG imager. In the first microfluidic configuration [60, 61], the biosensor array is microfabricated onto a glass substrate and has a footprint that is compatible to a standard 384-well microplate; and each biosensor has a dimension of 2 mm x 2 mm. The microfluidic device is made of PDMS using soft lithography, and contains a 4 x 4 array of microfluidic chambers, each having three inlets and one outlet (Figure 3a). The PDMS replica is attached to the biosensor array substrate to form a microchamber array covering 4 x 8 arrays of biosensors so only one biosensor is centered in each microchamber and the adjacent biosensor is sacrificed (Figure 3b). The flow is generated by syringe pumps connected to the inlets.

In the second microfluidic configuration [62], the microfluidic device has a footprint of conventional 96well microplates, and consists of 32 or 96 individually addressable microchambers, each containing two inlets and one outlet (Figure 3c). The two inlets locate at one side of a biosensor, opposite to the outlet. Such a T shape design ensures no mixing occurring when the two solutions flow side by side. The microchamber plate is made using injection overmolding of elastomer materials on top of the molded plastic part to ensure tight seal once the pins of the fluidic head are inserted to the device.

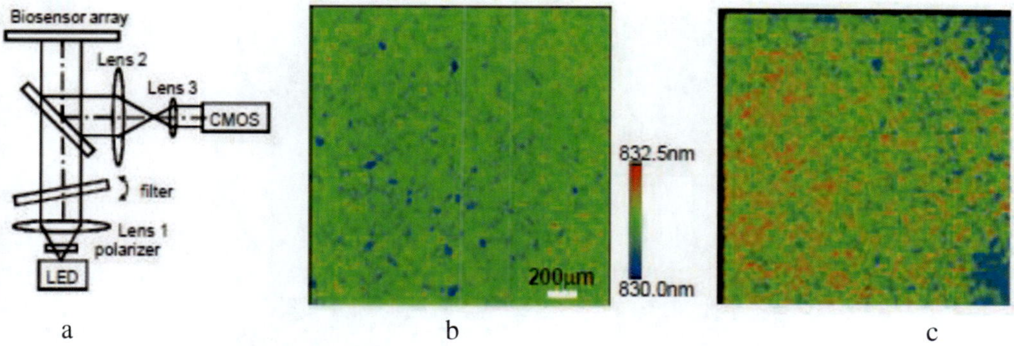

Figure 2. A RWG biosensor imager and its applications for cell signaling. (a) A schematic drawing showing the optical path of the imager. (b) A false-colored basal resonant wavelength images of a single biosensor bearing confluent A431 cells at 37°C. The image was obtained after the cell formed a monolayer, but before ligand stimulation. (c) A DMR image of the A431 2min after stimulation with 32nM epidermal growth factor (EGF), a natural agonist for EGF receptor endogenously expressed in A431 cells. The DMR image was obtained after the basal resonant wavelengths at all pixels were normalized to zero.

This system uses a fluidic head consisting of 96 pins to generate the flow using electronically controlled pumps. This system enables simultaneous measurements of cellular responses from the 4 x 8 biosensor array using the RWG imager. Because of the spatially resolved ability of the imager, each biosensor can be divided into two areas – the detection and referencing areas. When two solutions flow in parallel, the cellular responses arising from both areas can be compared (Figure 3d).

Other optical biosensors including whispering gallery mode (WGM) biosensors using high-Q resonators, zero-mode waveguide sensors, micro-ring resonators, and photonic bandgap biosensors have also been used to generate integrated optofluidic systems [26]. However, these opticfluidic systems have not been widely used for cell research, mostly due to limited throughput.

Microfluidic Cell Culture

One characteristic common to all organisms is the dynamic ability to coordinate constantly their activities with environmental changes. All cells produce and receive multiple, diverse signals. The function of communicating with the environment is achieved through distinct mechanisms including endocrinal, paracrinal, autocrinal and intracellular signal transduction.

In endocrinal signaling, the hormones formed are secreted from specialized endocrinal tissues into the extracellular medium and transported via the circulatory system to the target cells. In paracrinal signaling, the hormones secreted from the hormone producing cell reach the neighboring target cell via diffusion. In autocrinal signaling, the hormone secreted binds to and activates autocrine receptors on the same cell. These extracellular hormone signals are transduced into intracellular signaling that control many biochemical activities of cells and can also cause further release of signaling molecules.

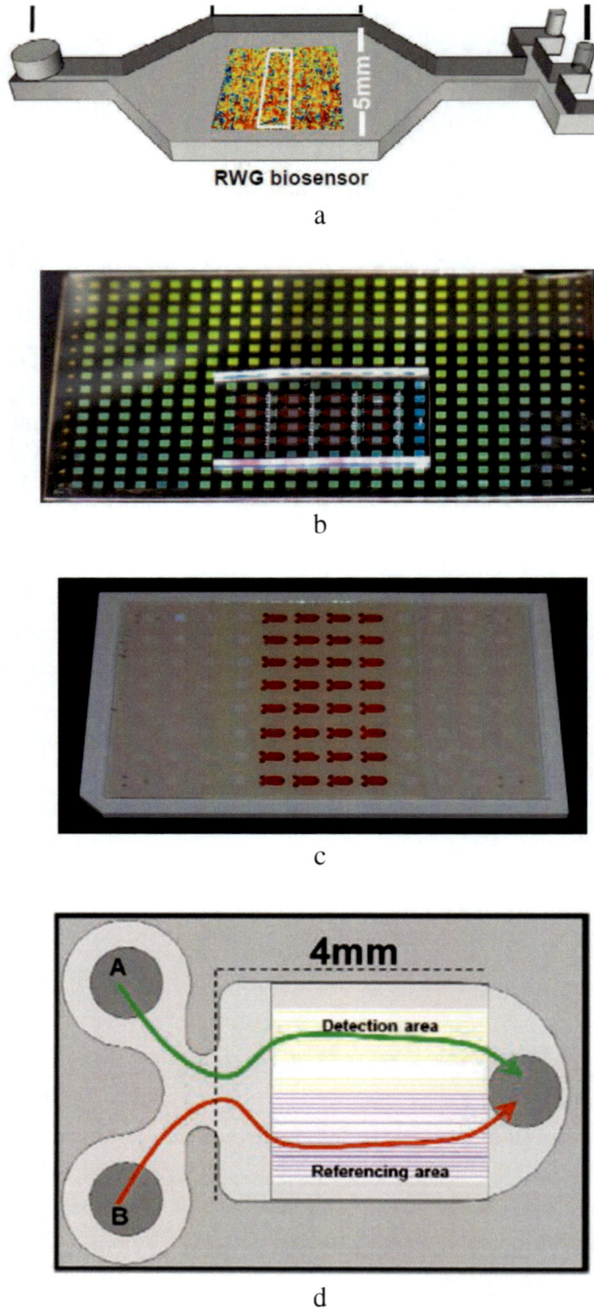

Figure 3. Configurations of microfluidic RWG devices for whole cell sensing. (a) A schematic showing the design of a microfluidic biosensor. The false colored image is an actual resonant image of a biosensor bearing a monolayer of A549 cells. The white box indicates the location at which the cellular responses were monitored. (b) A false colored image of a glass substrate having 384 biosensor array and a 4x4 microfluidic biosensor array within the substrate. The image was recorded using a digital camera by titling the biosensor plate under white light. (c) A bottom view photo of a 96well biosensor plate containing 4 x 8 functional microfluidic biosensors. A colored buffer solution was introduced to illustrate the microchamber geometry. (d) The geometry of the T-shaped microchamber and its side by side perfusion scheme. Produced from *Ref. 60* and *62* with permission.

There are two principal ways for cells to transduce external signals into intracellular pathways. In one way, the signal receipt and signal transduction occurs at the cell membrane by transmembrane receptors such as G protein-coupled receptors (GPCRs). In the other way, the messenger passes the cell membrane and binds to an intracellular receptor. Nonetheless, the binding of a hormone to its target receptor leads to the activation of the receptor, which, in turn, transmits the signal into complex biochemical cascades of kinases, proteases, adaptor proteins, transcription factors and so on, which act together to regulate cell physiology.

Microfluidic devices typically manipulate tiny volumes of fluids. Thus, the characteristics of microfluidic flow are important factors influencing cell cultures. At the microscale, fluid flow is governed by laminar flow regime due to the miniscule characteristic lengths of microfluidic channels. The laminar flow, whose convective flow profile is dependent on channel geometry, pressure drop, and fluid properties, allows precise calculation of mass transport as a function of time [63]. Laminar flow allows barrier-free adjacent perfusion of different reagents, which can be used for localized treatment of cells. Mass transport is directed by Brownian movement-driven diffusion [63]. Controlled diffusion and laminar flow in combination can be used to construct chemical gradients with high spatial resolution. Furthermore, within microfluidic channels the relative amount of surface area exposure experienced by a specific volume of fluid greatly increases, leading to faster diffusion and increased heat transfer [63]. However, the high surface-to-volume ratio could also present an issue related to potential absorption of certain molecules into the device material.

Conventional cell-based assays rely on static cell culture using microplates, where the culture medium is supplied in a batch-wise manner. Owing to its low manufacturing cost, easy to use, and automation compatibility, static cell culture has dominated in basic research and drug discovery where throughput is highly desired. However, static cell culture often encounters issues associated with demanding culture maintenance procedures, and inconsistency in the concentrations of growth factors, secreted molecules, metabolic byproducts, and oxygen due to respiratory and metabolic activity of cells. Highly reproducible static cell culture, a prerequisite for high throughput screening, can be challenging, in particular, when high density microplates are used.

Microfluidic perfusion culture is very attractive for several reasons. First, the perfusion culture can continuously provide nutrient supply and waste removal, resulting in relatively stable culture environment [25]. Second, the perfusion can provide additional functionalities including the generation of specific gradients of drug concentrations and of specific physical microenvironment (e.g. shear stress, interstitial fluid flow, or circulatory flow). Common to all signaling mechanisms is the presence of tightly regulated temporal and spatial gradients of signaling molecules and networks which encode the specificity of cellular responses [64]. The physical microenvironment including patterns of adhesive sites, fluid flow, substrate topography and stiffness could be designed into the device, and can influence the behavior and function of cells [65]. Tourovskaia et al. [66] presented a microfluidic perfusion system for studying the differentiation of myoblasts into myotubes. The device used a patterned substrate consisting of isolated lines of cell-adhesive coating, which are alternated with a robust cell-repellent coating, thus providing an *in vivo*-like spatial cues for muscle cell assembly. The confined fusion of myoblasts into aligned, isolated multinucleated myotubes was observed under the two-week microfluidic perfusion condition.

Common to cells in organisms is that cells typically reside in environments with very specific 3D features. However, most *in vitro* cell-based assays using adherent human cells are performed using two-dimensional (2D) cultures of cells onto plastic surfaces. These 2D cultured cells are amenable to both high throughput and high content screening, but fall short to mimic the cellular environments *in vivo*. Compared to standard 2D culture and bioreactors with large dimensions, microfluidic devices provide a much more effective means of controlling gradients of soluble signaling molecules, cell-cell contacts, and cell-extracellular matrix (ECM) interactions, thus more accurately mimicking the *in vivo* microenvironments [67]. The temporal and spatial gradients of soluble factors such as growth factors and oxygen can be introduced through microfluidic networks that allow distinct perfusion schemes [22]. Laminar flow in microfluidics can also be used to create desired mechanical strain, through shear, in the physiological range [68]. This is in particular relevant to many different types of cells and tissues including liver and kidney that are in close proximity with the microvascular network, wherein extensive perfusion is common.

The ECM onto which cells are harbored is part of environmental cues for regulating functions as diverse as tissue maintenance, immune response, and development [69]. The cell surface integrins often bind to ligands in the ECM substratum and transduce signals through their intracellular domains. The cells also sense the mechanical properties and organization of the ECM [70]. Cell signaling often leads to remodeling of its surrounding environment by producing or degrading ECM or soluble factors, and by exerting mechanical force [71]. Both compositions and organization of the ECM can regulate cell functions. The cell-cell and cell-matrix interactions can be precisely controlled using micropatterned 2D and/or 3D ECM structures, leading to ensembles of cells with defined geometry. The 2D ECM patterns, each with micrometer sizes can be generated using many surface patterning techniques such as microcontact printing and fluidic patterning [72]. The 3D scaffolds of ECM and other materials with feature sizes in the lower micrometre range can be generated using lamination, moulding and photo-polymerization techniques [73]. The patterning of ECMs and other biomaterials can encourage and direct cell behavior ranging from cell adhesion to gene expression to phenotype such as cell shape and organization [74]. Distinct spacing and geometry of a grid of ECM patches have been shown to have dramatic influence on cell behavior [75]. Both human and bovine capillary endothelial cells were found to switch from growth to apoptosis when the size of ECM adhesive islands is decreased to progressively restrict cell extension. Adhesive ECM patches also have been found to guide the orientation of the cell division axis [76], and to regulate fate decisions of human embryonic stems cells [77].

Co-culturing of multiple types of cells within a device is feasible, even under conditions that mimic normal 3D environmental and circulatory cues, so it is possible to create various artificial organ systems [24]. In a relatively simple configuration, two types of cells can culture together. For example, co-culture of hepatocytes with another helper cell line including liver epithelial cells and Kupffer cells can prolong the survival of cultured hepatocytes and help maintain liver-specific properties such as albumin secretion [78]. In more sophisticated configurations, microfluidic devices having multiple compartments that are interconnected via well-defined channels can be used to construct 3D human artificial organs consisting of more than one types of tissue cells [79]. These artificial organs simulate the physical, chemical, and physiological responses of natural organs and organ systems, and can be used as novel models of *in vitro* multicellular human organisms for the study of human physiology in an organ-specific context [80, 81], and drug safety in a pharmacokinetic-based manner [82, 83].

MICROFLUIDIC BIOSENSORS FOR ON-CHIP ELECTROPHYSIOLOGY ANALYSIS

The chemical-modulated electrophysiological activity is essential to the functions of excitable cells such as neurons, cardiomyocytes, muscle fibers and pancreatic beta cells. The electrophysiological measurements of ion channels, particularly cardiac human *Ether-à-go-go related gene* (hERG) potassium channel, are critical to drug discovery and development. The unintended and often promiscous inhibition of the hERG channel is a common cause for removal of therapeutic compounds from development and withdrawal of drugs from the market [84].

Glass micropipette-based patch clamping has been the golden standard for studying ion channel activities. In recent years, automated planar patch clamping has become the technique dominating drug discovery for ion channels and hERG-related drug safety assessment [85-87]. To record ion channel activities, these devices generally use a glass bottom microplate, each well having one or more micrometre-sized apertures in the planar glass substrate. The glass substrate separates two electrodes, and the micrometer-apertures are the recording sites. Microfluidics is used to locate cells onto apertures to form high-electrical-resistance seals, and also to present the cell with successive stimuli, so ion channel recording is possible. Common to these patch clamp techniques is the study of single or multiple ion channels in single cells.

Microelectrode arrays integrated with a microfluidic channel can be used for recording neuronal activity [88]. Recently, Tang *et al.* [89] developed a lateral patch-clamping device in a standard 1,536-well microplate format, and showed that its integration with robust microfluidics allows administering unique spatiotemporal doses of a drug to the cells. However, due to its unique configuration in which a silicon chip only contains twelve glass capillaries for cell attachment and clamping, only up to twelve captured cells can be studied in parallel.

Planar microelectrode arrays can also be integrated with microfluidics to measure the electrophysiological activities of ion channels. Surface patterning of the ECM materials can direct the attachment of neurons at precise points on the electrode arrays, and patterning can also be used to modulate the activity of the assembled neural networks [90, 91]. Werdich *et al.* [92] presented the use of a hybrid microfluidic/electrode device that can isolate and trap a single adult murine cardiomyocyte in a channel with recording electrodes, so the cardiomyocyte is confined to a volume of 100 pL, enabling a series of extracellular potential recordings.

Recently, Golden et al. [93] described a microfluidic patch clamping system, IonFlux, for automated electrophysiological recordings up to 64 samples in parallel. Compounds, cells, buffers and waste are deposited and handled in a single 96well or 384well microplate. Wells are interconnected by microfluidic channels, so cell trapping, whole cell recording, and compound perfusion can be accomplished using pneumatic pumps connected with the channels. Unlike conventional planar patch clamping devices which use micropores in a horizontal substrate [85-88], the IonFlux system uses patch sites that are microfrabricated into the sidewalls of the microfluidic channels. This system not only eliminates automated pipetting and fluidic tubing, and associated maintenance, but also allows for immediate replacement of compounds with buffer solution. Furthermore, unlike planar patch clamping devices which the temperature is not controlled, the IonFlux system provides accurate temperature control with a precision of 0.1°C using a heater to heat the microfluidic channels from room temperature to physiological temperature. This system was systematically validated with hERG channel stably expressed in Chinese hamster ovary cell line using a set of known hERG blockers.

MICROFLUIDIC BIOSENSORS FOR SENSING CELL MECHANICS

Living organisms also face diverse biomechanical environments, beside complex biochemical environments [94]. Cells perform a multitude of complex mechanical functions including migration, spreading, contracting, and relaxation, in response to both chemical and physical cues. Both internal and external physical cues can be converted into biochemical signals and integrated into distinct cellular responses; and chemical cues can also produce physical forces in cells [95]. Central to cell mechanics is cytoskeleton, a dynamic and adaptive structure interconnected network of filamentous polymers and regulatory proteins [96, 97]. Cytoskeleton is essential to the spatial organization of cellular contents, the physical and biochemical connections of cells with the external environment, and many cellular responses by coordinating forces that enable the cell to move and change shape.

Combining microfluidics with advanced microscopy and labeling techniques have greatly advanced our understanding of cell mechanics [95, 98]. These studies have revealed distinct aspects of cell mechanics, such as the altered deformability of cells during cancer development, and the altered adhesion, gene expression, morphology, signaling pathways, and growth of many types of cells under shear flow stress [98-100]. Microfluidics, coupled with advanced imaging techniques, can provide insights for the mechanics of tissue expansion. Serra-Picamal et al. [101] reported an innovative approach combining soft lithography, traction force microscopy and monolayer stress microscopy for probing the mechanical rules governing collective migration during tissue expansion. MDCK cells are seeded into the rectangular opening of a PDMS membrane deposited on a collagen I-coated polyacrylamide gel. After cells reach confluence and relatively high cell density, the monolayer sheet exhibited features typical of a stable epithelium. The PDMS membrane is then removed, leading to collective cellular migration towards newly available free space. Over the course of 10 hours the total area of the monolayer sheet more than doubled, with less than half of the total area being attributable to proliferation and the rest being attributable to cell spreading and flattening. High resolution imaging reveals that stress propagates in a wave, from the moving front to the center of the cell group, and suggests a mechanism of long-range cell guidance, symmetry breaking and pattern formation during monolayer expansion.

Microfluidic biosensors should provide an attractive alternative to understand the molecular, physical, cellular, and systems mechanisms of cell mechanics. The biosensors including RWG and electric biosensors can measure non-invasively the mechanical responses of cells under shear stress. Certain biosensors such as AFM can also directly produce an external force to a single cell. Other biosensors such as SAW can be used to manipulate and pattern cells [36], and could also deliver droplets of chemicals to specific cells [102]. Himmelhaus and Francois [103] presented the use of WGM biosensor to directly measure the biomechanical stress of a live cell during endocytosis. The fluorescent microspheres are used as microscopic remote optical sensors to excite distinct whispering gallery modes, which are then used as a transducer mechanism. Monitoring of the WGMs throughout the uptake of the microsphere into the cell enabled the determination of the deformation experienced by the microsphere, characterized by both a broadening and a blue shift of the resonances, and consequently the stress induced by the cell. The results reveal an unexpectedly high stress with a magnitude of up to five times that of the passive cortical tension, which can be only explained by unknown active stress component(s) induced by the cytoskeletal machinery during particle uptake.

MICROFLUIDIC BIOSENSORS FOR PROBING CELLULAR RESPONSES AND DRUG ACTIONS UNDER CONTROLLED CHEMICAL ENVIRONMENTS

In the open systems of *in vivo* experiments, both receptor signaling and drug actions are often far distant from equilibrium conditions. For receptor signaling involving endogenous signaling molecules, these molecules are secreted from a signaling-releasing cell and often present complex spatial and temporal gradients to their target cells [104]. In some cases, *in vivo* receptor functions are often dependent on very short, transient stimuli that may not allow the achievement of a steady state [105]. For *in vivo* drug actions, exogenous drug molecules may reach their target cells with distinct temporal concentration gradients, due to dynamic flow and drug metabolism in the human body [14]. The duration and gradients of chemical stimuli may have great influence on receptor signaling and cellular responses.

However, most *in vitro* cell-based assays use a static stimulation scheme to probe the biological effects of drug molecules using standard microplates. Under such a condition a compound once introduced remains constant in concentration and is exposed to the cells throughout the assays. Furthermore, *in vitro* receptor signaling is generally analyzed under equilibrium conditions, and drug actions are assessed using end-point assays. Thus, there is strong need in pharmacological assays that can manifest the impact of the duration and gradients of drug molecules on receptor signaling and cellular responses in real time.

Microfluidic biosensors have great advantages for investigating receptor signaling and drug actions. Label-free biosensors are capable of non-invasively and real-time monitoring whole cell phenotypic responses with wide pathway coverage, so receptor signaling and drug actions under different conditions can be assessed at the whole cell level. Microfluidics allows generating chemical signals with high temporal and spatial resolution. Several microfluidic devices have been demonstrated to allow rapid solution exchange down to a few milliseconds using several different techniques including interface shifting, so it is possible to probe rapid receptor signaling and drug actions [106]. Recently, Sun *et al.* showed that hydrodynamic gated injection can be used to generate a series of 100 millisecond chemical pulse with a 50sec interval to study the signaling of endogenous purinergic P2Y receptors in HeLa cells [106]. Microfluidics also can alter and minimize autocrine/paracrine signaling, thus making it possible to introduce exogenous factors in a controllable manner so the exact cellular mechanisms for cancer development and stem cell maintenance and differentiation can be investigated [107].

Receptor signaling is known to consist of a series of spatial and temporal events, each of which has distinct characteristics in kinetics, dynamics, amplitude and duration [64, 108]. Therefore, it is highly possible that rapid receptor events can occur under very short agonist stimulation condition, whereas certain slow and persistent receptor events require continuous receptor occupancy by the agonist [60, 61, 109].

Recently, we investigated the impact of receptor signaling and drug actions under different stimulation schemes using a microfluidic RWG biosensor system, in comparison with microplate-based RWG assays [60, 61]. A PDMS microfluidic device connected to pumps was used to generate a pulse of several agonists including epinephrine and salmeterol to activate the endogenous β_2-adrenerigc receptor (β_2-AR) in A431 cells for short times, and the receptor activation-induced DMR signal was monitored in real time before and after the pulse stimulation [61]. Results showed that the short pulse (1 or 2min) stimulation with 2nM epinephrine led to a DMR signal distinct from the DMR arising from continuous stimulation under both static and shear conditions – the pulse stimulation split the single positive-DMR induced by the continuous epinephrine stimulation into a biphasic event (comparing Figure4a with Figure 4b). Furthermore, receptor internalization and ERK phosphorylation were also found to be dependent on the duration of agonist exposure. Under the 2min pulse stimulation, only small numbers of receptors were internalized, and the ERK phosphorylation was more sustained than that under continuous stimulation. Similarly, the DMR signals arising from the protease activated receptor peptide agonist SFLLR-amide in A549 was also found to be dependent on the duration of the agonist exposure [60]. These results suggest for the first time that receptor signaling not only at distinct signaling molecule level, but also at the whole cell level, is sensitive to the duration of agonist exposure. Previous fluorescent imaging studies have showed that distinct cellular events downstream the receptor activation have different sensitivity to the receptor occupancy by the agonists. For β_2-adrenergic receptor (β_2-AR) Violin et al. [110] found that the β-arrestin recruitment, possibly β-arrestin-mediated signaling, downstream the receptor activation, requires persistent ligand occupancy. Lohse and his colleagues found that in thyroid cells the endogenous thyroid-stimulating hormone (TSH) receptor can continue stimulating cAMP production after co-internalization with TSH; and cAMP signaling by internalized receptors is persistent, whereas that by cell-surface receptors is apparently transient [111, 112].

Drug binding is the prerequisite step to all drug-target interactions. However, the physiologically relevant result is the functional consequence of drug binding. Studies of the biochemical efficiency of marketed drugs has suggested that non-equilibrium mechanisms of action may lead to efficacious drugs [113], and increased drug residence time (the reciprocal of K_{off}) is a more critical indicator for the *in vivo* effects of drugs than their *in vitro* equilibrium constants [28, 114]. Using the microfluidic RWG biosensor we had found that the DMR arising from different β_2-AR agonists exhibited different sensitivities to agonist exposure duration and removal [61]. Under the 2min pulse stimulation conditions, salmeterol resulted in a DMR identical to those obtained under static conditions (Figure 4c). The ligand-dependent sensitivity to the agonist exposure duration and removal is consistent with the fact that epinephrine is a reversible agonist with a fast off rate, but salmeterol is a long-lasting agonist with much slower off rate. Nishijima et al. [115] used SPR under microfluidics to investigate the impact of a short duration treatment with anti-cancer drugs on cell apoptosis. Results showed that a short-term (~1hr) treatment of liver cell line HepG2 with roscovitine (a cell cycle arrest drug) can have long-term (~48 hours) effects on cell viability.

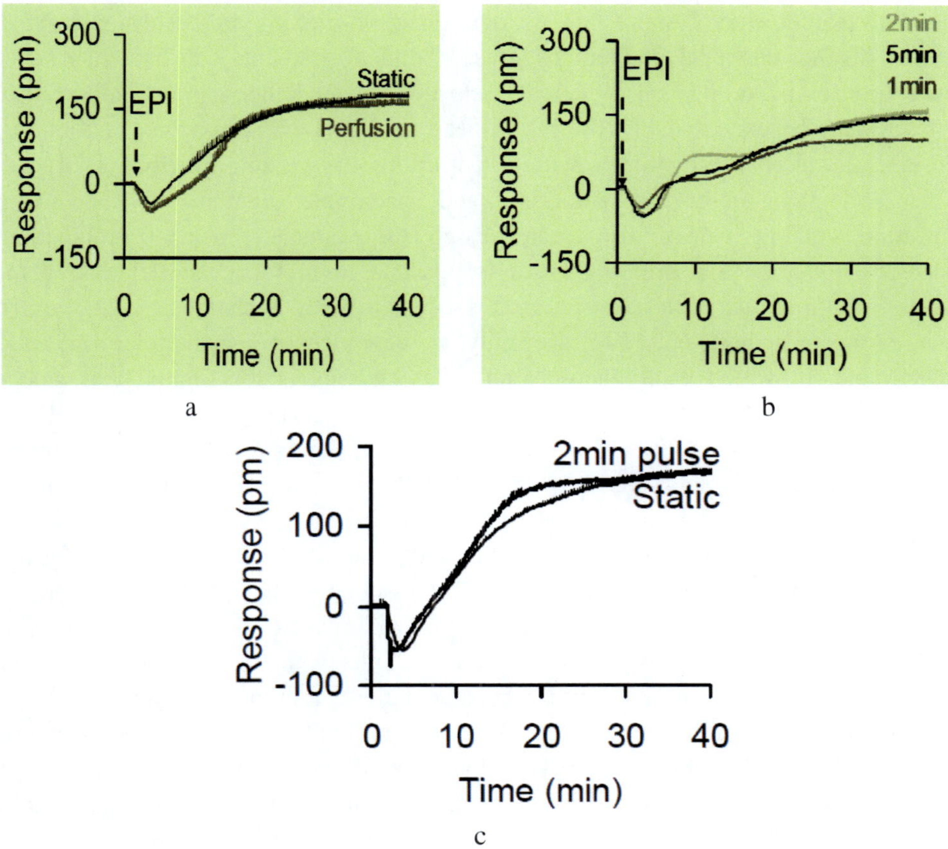

Figure 4. The DMR signals of A431 cells induced by distinct β_2AR agonists under different conditions. (a) The DMR of 2nM epinephrine (EPI) obtained using microplate-based static condition as well as continuous perfusion condition. (b) The DMR of 2nM epinephrine (EPI) obtained using the pulse perfusion scheme. The pulse or agonist exposure time was 1 min, 2 min or 5 min. (c) The DMR of 100nM salmeterol under the 2min pulse stimulation in comparison with that obtained under the continuous perfusion condition. The perfusion rate was 1µl/min for all. Data represents mean±s.d. from 3 independent measurements. Produced from *Ref. 61* with permission.

Microfluidics not only can control the duration of agonist exposure, but also control the timing of the functional recovery of desensitized receptors [60-62]. Prolonged or repeated use of agonist drugs often leads to loss of their effects and causes tolerance, a pervasive phenomenon termed tachyphylaxis, refractoriness, or desensitization [116-118]. Many molecular aspects of receptor desensitization and resensitization have been discovered in the past decades; however, understanding of these processes at the whole cell and systems level still lags behind. Using microfluidic RWG biosensors to control the timing of functional recovery of desensitized receptors, we had found that distinct cellular mechanisms are accountable for the ligand-directed desensitization of the β2-AR [61]; receptor recycling and synthesis are essential to ligand-directed resensitization after the PAR1 is desensitized [60]; and there is a clear difference between the agonist-induced homologous desensitization and the downstream activator forskolin induced heterologous desensitization of the β_2AR [62].

Microfluidic devices can be used to generate controlled perturbation of the cellular environment in time and space for adherent cells by controlling flow over the cells [22]. The laminar flow under microfluidics has unique diffusive mixing properties, allowing the creation of distinct chemical gradient patterns. Chemical gradients can encourage specific cellular responses including cell directional sensing that are otherwise difficult to achieve using microplate-based assays [119-121]. Linear gradients of exogenous factors can be created in static microfluidic systems without perturbing the existing distribution of secreted molecules, thereby preserving autocrine and paracrine signaling [122]. Complex concentration gradients can be also created using, for example, repeated combinations of flow-stream lamination and splitting, so it is possible to perform parallel dose response analysis of multiple spatially resolved groups of cells with a chamber [123], and even multiple regions of a single cell [124]. Recently, we developed a T shape microfluidic RWB biosensor plate, in which each microchamber has two inlets locating at one side of a biosensor, opposite to the one outlet [62]. The T shape design ensures minimal mixing occurring when the two solutions flow side by side, so it is possible to perform referencing studies when two distinct solutions flow in parallel using the RWG imaging system (Figure 3d).

Microfluidic electric biosensors can also be used to detect cell phenotypic changes under different fluid conditions. Primiceri et al. [125] designed a PDMS-based, microporous membrane separated transwell device for cell migration studies. Cells are seeded in the upper compartment through microfluidic channels. Upon exposure to distinct ECM components including collagen IV and laminin I and V, hepatocellular carcinoma cells migrate across the pores of the membrane and land on an electric biosensor located in the surface of the bottom compartment. Xu et al. [126] described a PDMS-based microfluidic device for real-time monitoring of drug-induced cytotoxicity using electric biosensor. A gradient generator was used to present multiple doses of the anticancer drug cisplatin at the same time, and the apoptotic-related responses of four different cancer cell lines were detected in real time using impedance-sensing electrodes. Grist et al. [127] presented a microfluidic device for hypoxia study. Hypoxia is a characteristic feature of locally advanced solid tumors. Two control channels were designed to be adjacent to the cell culture chamber and separated by thin PDMS barriers. Oxygenated and deoxygenated buffers were flowed through these control channels to form oxygen gradient across the cell chamber due to the gas permeability of PDMS barriers. A sensor containing oxygen-sensitive platinum(II) octaethylporphyrin ketone was used to detect oxygen content and gradients. The effects of rapamycin and tamoxifen on cell viability and autophagy after 3 days treatment were studied using fluorescent imaging.

Microfluidic Biosensors for Single Cell Analysis

Current assays dominating in cell research and drug discovery are mostly dealt with an averaged response directly from a cell population or indirectly from a population of single cells, driven by considerations associated robustness and operational simplicity [128]. However, cells are extremely plastic and dynamic, and are almost always heterogeneous in function and fate. Cellular heterogeneity or cell-to-cell variability was long recognized using cell-population-averaging and single cell techniques. As a matter of fact, robust screening using frozen cells after certain synchronization procedures such as cell cycle arrest is an industrial routine to minimize assay variability during drug discovery process [129]. Single cell measurements and lineage-tracing experiments have suggested that heterogeneity is a characteristic intrinsic to many cell signaling and cell-fate decision processes including cell division, death and differentiation (or dedifferentiation) [128, 130-133]. Intrinsic stochastic events, which are associated with small numbers of molecules and gene copies, generate 'noise' [134]. This noise might need to be minimized by the cell, but it might also be used as part of the mechanism of decision-making [135]. To understand the plasticity of cells, it is vital to measure quantitatively and dynamically the molecular processes that underlie cell-fate decisions in single cells.

Single-cell experiments reveal heterogeneities in cell properties that are otherwise obscured by population-averaged measurements. Microfluidic technologies have become an indispensable component for studying cell-to-cell variability. Integrated microfluidic systems can be used to trap, manipulate and analyze single cells so heterogeneity in proteins, gene expression, cell signaling, and phenotypic dynamics can be analyzed at the single cell level [136-143].

For single cell analysis, it is critical to trap and manipulate the cells to achieve reliable culture of single cell with controllable cell–cell interactions and microenvironments. Di Carlo et al. [144] presented a microfluidic based dynamic single cell culture array using a trapping array consisting of physical U-shaped hydrodynamic trapping structures with geometries that are biased to trap only single cells. The arrays within the chambers consist of U-shaped PDMS structures that are 40 μm in height and were offset from the substrate by 2 μm. Cell loading was done in less than 30 seconds, leading to greater than 90% of cells being adherent.

Owing of its non-invasive nature, microfluidic biosensors can real-time monitor the cellular responses *in situ* through the whole life cycle of cells at the single cell level. Ghenim et al. [145] presented a microfluidic biosensor device to real-time monitor the motility and mitosis of a single cell adherent on an electric biosensor. The biosensor used a tiny gold electrode with a diameter of 20 μm, a dimension close to a HeLa cell. The electrode can be switched electrochemically to favor cell adhesion, and single cells were attracted to the electrode using positive dielectrophoresis. The impedance measurement allows tracking the entire life cycle of a single cell including cell adhesion, cell motion and division.

Cells in suspension are usually transported with flow. However, cells can be physically retained in the devices by filters or traps. By combining concentration gradients and flow of cells with hydrodynamic traps that retain cells in a fixed position, microfluidic biosensors could have high resolution to measure specific phenotypes of single cells. Hua and Pennell [146] reported a microfluidic biosensor device comprising a narrow sensing region with a chevron-like feature created along the fluidic path, and its use for measuring the volume of single cells. The chevron-like feature was designed to capture single suspended MDCK cells in the solution. The changes in cell volume induced by hypotonic challenges and drugs were measured in situ using electric impedance biosensor. Results showed that the maximum swelling and the time course of swelling vary between individual cells following hypotonic stimulation.

Su et al. [147] described a microfluidic flow cytometric technique to study the distribution of nanometer-size mitochondria in single cells. This technique was based on 2D light scattering imaging of single malignant lymphoid Jurkat cells and normal hematopoietic stem cells (cord blood CD34+ cells) trapped within a three layer microfluidic cavity. Results showed that the mitochondria were randomly distributed in the Jurkat cells, but aggregated in the CD34+ cell.

Microfluidic biosensors can be used to measure the biophysical (mechanical and electrical) properties of single cells at high throughput. Microchannels can be used to generate fluid stress for investigating membrane viscoelastic properties of cells. Deformation index can be determined using high-speed imaging when single cells are forced to pass through a constriction channel. The cell deformability can be used as an intrinsic biophysical marker to classify a specific population of cells from patient samples, and to decode the (de)differentiation of stem cells. For example, human red blood cells (RBCs) are known to have exceptional capacity to deform, a property critical to both macro and microcirculation. The impaired deformability of RBCs has been linked to a range of diseases such as sepsis, malaria, sickle cell anemia (hemoglobin disorder), and myocardial ischemia and microvascular dysfunction. Zheng et al. [148] described the use of a microfluidic electric biosensor device for measurements of RBCs at a speed of 100–150 cells per second. Electrical impedance measurements were made when single RBCs flow through a constriction channel that is marginally smaller than RBCs' diameters. Multi-parameter analysis of 84,073 adult RBCs (from 5 adult blood samples) and 82,253 neonatal RBCs (from 5 newborn blood samples) revealed different biophysical properties across samples and between the adult and neonatal RBC populations.

Ang et al. [149] developed a graphene transistor array integrated with microfluidic flow cytometry for the "flow-catch-release" sensing of malaria-infected red blood cells at the single-cell level. Graphene was functionalized with the endothelial CD36 receptors for the selective capture of the malaria-infected cells when the diseased blood flows through the microfluidic channel.

Malaria-infected red blood cells were found to induce highly sensitive, capacitively coupled changes in the conductivity of graphene. Microfluidic electric impedance devices have also been used to characterize the deformability of several other types of cells including human breast cancer cells [150, 151], malignant leukocytes [152], and osteoblasts and osteocytes [153].

Cell signaling is encoded with a series of spatial and temporal gradients of biomolecules and their network interactions. Current single cell analysis is mostly focused on specific sets of molecules (e.g., proteins, genes), but with limited information on the integration of signaling components at the whole cell level. Using a high resolution RWG imager that is capable of resolving cell signaling at the single cell level, we found that in quiescent human epidermal carcinoma A431 cells at 37°C/5% CO_2 the DMR signal arising from epidermal growth factor (EGF) is quite diverse at the single cell level [48]. The EGF DMR consisted of three phases – an initial rapid positive-DMR, followed by a decayed negative-DMR, and a small recovery positive-DMR (Figure 5a).

The initial rapid positive EGF response associated with the protein kinase C (PKC) pathway exhibited a distribution having two peaks centered at 100 pm and 600 pm, respectively (Figure 5b) [53], suggesting the possible presence of two distinct quiescent states achieved through distinct mechanisms. Different quiescent states may have different basal activity of PKC pathway. However, the signals at both 9min and 60min poststimulation were relatively uniform (Figure 5c), suggesting that regardless of the cellular status the late EGF responses are less variable.

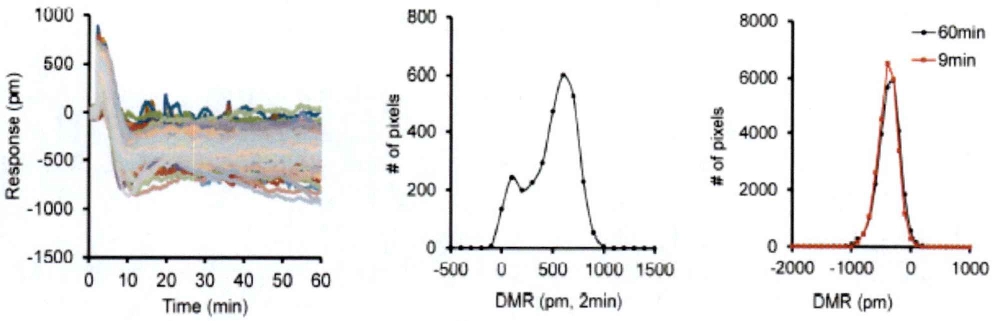

a b c

Figure 5. Cell-to-cell variability of 32nM EGF induced DMR in quiescent A431 cells at the single cell level. (a) The real time kinetic DMR at 166 pixels across a biosensor bearing monolayer A431 cells. (b) Distribution of the EGF responses at 2 min poststimulation. (c) Distribution of the EGF responses at 9min poststimulation. The bin size was 100 pm for all histograms. Produced from *Ref. 48* with permission.

CONCLUSION

Label-free biosensors for whole cell sensing come with greater advantages, compared to conventional molecular assays. Instead of being restrictive to the specific molecular mechanism of action defined by conventional assays which are usually proximal to the receptor, label-free whole cell assays provide phenotypic measurements of receptor biology and drug pharmacology with wide pathway coverage [13, 154-157]. This is significant since many ligands have multiple efficacies and also can differentially activate selective pathways among complex signaling networks of a receptor [158-160]. Label-free whole cell assays have high sensitivity, so native cells can be directly assayed without any labels or genetic manipulations. The use of native cells allows preservation of the relative stoichiometry of receptors and cellular signaling components, making label-free whole cell assays valuable for assessing drug activity in different cells types including disease relevant cells. Label-free whole cell assays allow the observation of drug effect on complete cellular systems in real time, enabling multiparameters analysis based on real time kinetics for decoding multiplex signaling pathways [13, 154, 161].

In recent years, microfluidics has been gaining momentum in cell research, particularly single-cell analysis, for several reasons. First, microfluidics allows precise manipulation and control of single cells, so it is possible to present single cells for analysis, as well as to establish functional culture of cells in an environment that is either close to or distant from *in vivo* environments. Second, micro-scale fluid flow enables precise and high-resolution microenvironment control, so cellular responses under different spatiotemporal molecular gradients can be studied with single-cell resolution. Microfluidic biosensor systems have been explored and used for on-chip electrophysiology analysis, cell mechanics sensing, cell biology and drug pharmacology studies under controlled chemical environments, and single cell manipulation and analysis. The integration of label-free biosensor with microfluidics opens new possibility in both basic research and drug discovery, in particular, in the area where cellular heterogeneity and phenotypic signatures are important.

REFERENCES

[1] Stockwell, B. R. (2000). Chemical genetics: ligand-based discovery of gene function. *Nat. Rev. Genet.*, 1, 116-125.

[2] Hopkins, A. L. and Groom, C. R. (2002). The druggable genome. *Nat. Rev. Drug Discov.*, 1, 727-30.

[3] Hannon, G. J. (2002). RNA interference. *Nature*, 418, 244–251.

[4] Santiago, Y., Chan, E., Liu, P. Q., Orlando, S., Zhang, L., Urnov, F. D., Holmes, M. C., Guschin, D., Waite, A., Miller, J. C., Rebar, E. J., Gregory, P. D., Klug, A., and Collingwood, T. N. (2008). Targeted gene knockout in mammalian cells by using engineered zinc-finger nucleases. *Proc. Natl. Acad. Sci. US.,* 105, 5809-5814.

[5] Hannig, G. and Makrides, S. (1998). Strategies for optimizing heterologous protein expression in *Escherichia coli. Trends Biotechnol.*, 16, 54–60.

[6] Andrianantoandro, E., Basu, S., Karig, D. K., and Weiss, R. (2006). Synthetic biology: new engineering rules for an emerging discipline. *Mol. Syst. Biol.*, 2, 2006.0028.

[7] Kunkel, T. A. (1985). Rapid and efficient site-specific mutagenesis without phenotypic selection. *Proc. Natl. Acad. Sci. US.*, 82, 488-492.

[8] Sako, Y., Minoghchi, S., and Yanagida, T. (2000). Single-molecule imaging of EGFR signalling on the surface of living cells. *Nat. Cell Biol.,* 2,168-172.

[9] Rust, M. J., Bates, M., and Zhuang, X. (2006). Sub-diffraction-limit imaging by stochastic optical reconstruction microscopy (STORM). *Nat. Meth.*, 3, 793-795.

[10] Gupta, P. B., Onder, T. T., Jiang, G., Tao, K., Kuperwasser, C., Weinberg, R. A., and Lander, E. S. (2009). Identification of selective inhibitors of cancer stem cells by high-throughput screening. *Cell*, 138, 645-659.

[11] Brouzes, E., Medkova, M., Savenelli, N., Marran, D., Twardowski, M., Hutchison, J. B., Rothberg, J. M., Link, D. R., Perrimon, N., and Samuels, M. L. (2009). Droplet microfluidic technology for single-cell high-throughput screening. *Proc. Natl. Acad. Sci. USA*, 106, 14195-14200.

[12] Swinney, D. C. and Anthony, J. (2011). How were new medicines discovered? *Nat. Rev. Drug Discov.*, 10, 507-519.

[13] Fang, Y. (2011). The development of label-free cellular assays for drug discovery. *Exp. Opin. Drug Discov.*, 6, 1285-1298.

[14] Gleeson, M., Hersey, A., Montanari, D., and Overington, J. Probing the links between in vitro potency, ADMET and physicochemical parameters. *Nat. Rev. Drug Discov.*, 2011, 10, 197-208.

[15] Butcher, E. (2005). Can cell systems biology rescue drug discovery? *Nat. Rev. Drug Discov.*, 4, 461-467.

[16] Kenakin, T. (2009). Cellular assays as portals to seven-transmembrane receptor-based drug discovery. *Nat. Rev. Drug Discov.*, 8, 617-626.

[17] Dittrich, P. S. and Manz, A. (2006). Lab-on-a-chip: microfluidics in drug discovery. *Nat. Rev. Drug Discov.*, 5, 210-218.

[18] Kang, L., Chung, B. G., Langer, R., and Khademhosseini, A. (2008). Microfluidics for drug discovery and development: from target selection to product lifecycle management. *Drug Discov. Today*, 13, 1-13.

[19] Wootton, R. C. and Demello, A. J. (2012). Microfluidics: Analog-to-digital drug screening. *Nature*, 483, 43-44.

[20] Neuži, P., Giselbrecht, S., Länge, K., Huang, T. J., and Manz, A. (2012). Revisiting lab-on-a-chip technology for drug discovery. *Nat. Rev. Drug Discov.*, 11, 620-632.

[21] Whitesides, G. M. (2006). The origins and the future of microfluidics. *Nature*, 442, 368-373.

[22] Keenan, T. M. and Folch, A. (2008). Biomolecular gradients in cell culture systems. *Lab Chip*, 8, 34-57.

[23] Ei-Ali, J., Sorger, P. K. and Jensen, K. F. (2006). Cells on chips. *Nature*, 442, 403-411.

[24] Khademhosseini, A., Langer, R., Borenstein, J., and Vacanti, J.P. (2006). Microscale technologies for tissue engineering and biology. *Proc. Natl. Acad. Sci. USA,* 103, 2480-2487.

[25] Wu, M.-H., Huang, S.-B. and Lee, G.-B. (2010). Microfluidic cell culture systems for drug research. *Lab Chip*, 10, 939-956.

[26] Psaltis, D., Quake, S. R. and Yang, C. (2006). Developing optofluidic technology through the fusion of microfluidics and optics. *Nature*, 442, 381-386.

[27] Fritzsch, F. S., Dusny, C., Frick. O., and Schmid, A. (2012). Single-cell analysis in biotechnology, systems biology, and biocatalysis. *Annu. Rev. Chem. Biomol. Eng.*, 3,129-55.

[28] Fang, Y. (2012). Receptor-ligand interaction platforms and their applications for drug discovery. *Exp. Opin. Drug Discov.,* doi:10.1517/17460441.2012.715631.

[29] McGuinness, R. (2007). Impedance-based cellular assay technologies: recent advances, future promise. *Curr. Opin. Pharmacol.* 7, 535–540.

[30] A-Hassan, E., Heinz, W. F., Antonik, M. D., D'Costa, N. P., Nageswaran, S., Schoenenberger, C.-A., and Hoh, J. H. (1998). Relative microelastic mapping of living cells by atomic force microscopy. *Biophys. J.,* 74, 1564-1578.

[31] Ehrlicher, A. and Hartwig, J. H. (2011). Cell mechanics: contracting to stiffness. *Nat. Materials*, 10, 12-13.

[32] Cheng, J.-X. and Xie, X. S. (2004). Coherent anti-Stokes Raman scattering microscopy: instrumentation, theory, and applications. *J. Phys. Chem. B*, 108, 827-840.

[33] Wang, M. C., Min, W., Freudiger, C. W., Ruvkun, G., and Xie, X. S. (2011). RNAi screening for fat regulatory genes with SRS microscopy. *Nat. Meth.*, 8, 135-138.

[34] Beuthan J, Minet O, Helfmann J, Herrig, M., and Müller, G. (1996). The spatial variation of the refractive index in biological cells. *Phys. Med. Biol.*, 41, 369-382.

[35] Chen, J. Y., Li, M., Penn, L. S., and Xi, J. (2011). Real-time and label-free detection of cellular response to signaling mediated by distinct subclasses of epidermal growth factor receptors. *Anal. Chem.*, 83, 3141-3146.

[36] Ding, X., Lin, S. C. S., Kiraly, B., Yue, H., Li, S., Shi, J., Benkovic, S. J., and Huang, T. J. (2012). On-chip manipulation of single microparticles, cells, and organisms using surface acoustic waves. *Proc. Natl. Acad. Sci. USA,* 109, 11105-11109.

[37] Ding, X., Shi, J., Lin, S. C. S., Yazdi, S., Kiraly, B., and Huang, T. J. (2012). Tunable patterning of microparticles and cells using standing surface acoustic waves. *Lab Chip*, 12, 2491-2497.

[38] Kang, H. W., Muramatsu, H., Lee, B. J., and Kwon, Y. S. (2010). Monitoring of anticancer effect of cisplatin and 5-fluorouracil on HepG2 cells by quartz crystal microbalance and micro CCD camera. *Biosens. Bioelectron.*, 26, 1576-1581.

[39] Fang, Y., Ferrie, A. M. and Li, G. (2005). Probing cytoskeleton modulation by optical biosensors. *FEBS Lett.*, 579, 4175-4180.

[40] Fang, Y., Ferrie, A. M., Fontaine, N. H., and Yuen, P. K. (2005). Characteristics of dynamic mass redistribution of EGF receptor signaling in living cells measured with label free optical biosensors. *Anal. Chem.,* 77, 5720-5725.

[41] Fang, Y., Li, G. and Peng, J. (2005). Optical biosensor provides insights for bradykinin B2 receptor signaling in A431 cells. *FEBS Lett.,* 2005, 579, 6365-6374.

[42] Fang, Y., Ferrie, A. M., Fontaine, N. H., Mauro, J., and Balakrishnan, J. (2006). Resonant waveguide grating biosensor for living cell sensing. *Biophys. J.,* 91, 1925-1940.

[43] Tanaka, M., Hiragun, T., Tsutsui, T., Yanase, Y., Suzuki, H., and Hide, M. (2008). Surface plasmon resonance biosensor detects the downstream events of active PKCβ in antigen-stimulated mast cells. *Biosens. Bioelectron.,* 23, 1652-1658.

[44] Ziblat, R., Lirtsman, V., Davidov, D., and Aroetim B. (2006). Infrared surface plasmon resonance: a novel tool for real time sensing of variations in living cells. *Biophys. J.*, 90, 2592–2599.

[45] Fang, Y., Frutos, A. G. and Verklereen, R. (2008). Label-free cell assays for GPCR screening. *Comb. Chem. High Throughput Screen.*, 11, 357-369.

[46] Suraniti, E., Sollier, E., Calemczuk, R., Livache, T., Marche, P. N., Villiers, M. B., and Roupioz, Y. (2007). Real-time detection of lymphocytes binding on an antibody chip using SPR imaging. *Lab Chip*, 7, 1206-1208.

[47] Ferrie, A. M., Wu, Q. and Fang, Y. (2010). Resonant waveguide grating imager for live cell sensing. *Appl. Phys. Lett.*, 97, 223704.

[48] Ferrie, A. M., Deichmann, O. D., Wu, Q., and Fang, Y. (2012). High resolution resonant waveguide grating imager for cell cluster analysis under physiological condition. *Appl. Phys. Lett.*, 100, 223701.

[49] Zhang, P. C., Bai, C., Huang, Y. M., Zhao, H., Fang, Y., Wang, N. X., and Li, Q. (1995). Atomic force microscopy study of fine structures of the entire surface of red blood cells. *Scanning Microsc.*, 9, 981-989.

[50] Fang, Y. (2011). Label-free biosensors for cell biology. *Intl. J. Electrochem.,* 2011, 460850.

[51] Pai, S., Verrier, F., Sun, H., Hu, H., Ferrie, A. M., Eshraghi, A., and Fang, Y. (2012). Dynamic mass redistribution assay decodes differentiation of a neural progenitor stem cell. *J. Biomol. Screen.*, 17, DOI: 10.1177/1087057112455059.

[52] Kenakin, T. P. (2010). A holistic view of GPCR signaling. *Nat. Biotechnol.,* 28, 928–929.

[53] Fang, Y. (2010). Probing cancer signaling with resonant waveguide grating biosensors. *Exp. Opin. Drug Discov.*, 5, 1237–1248.

[54] Xia, Y. and Whitesides, G. M. (1998). Soft Lithography. *Angew. Chem. Int. Ed. Engl.*, 37, 551–575.

[55] Kim, S. M., Lee, S. H. and Suh, K. Y. (2008). Cell research with physically modified microfluidic channels: a review. *Lab Chip*, 8, 1015-1023.

[56] Mukhopadhyay, R. (2007). When PDMS isn't the best. *Anal. Chem.*, 79, 3248-3253.

[57] Vozzi G., Flaim C., Ahluwalia A., and Bhatia S. (2003). Fabrication of PLGA scaffolds using soft lithography and microsyringe deposition. *Biomaterials*, 24, 2533–2540.

[58] Wang, Y., Ameer, G. A., Sheppard, B. J., and Langer, R. (2002). A tough biodegradable elastomer. *Nat. Biotechnol.*, 20, 602–606.

[59] Choi, C. J. and Cunningham, B. T. (2007). A 96-well microplate incorporating a replica molded microfluidic network integrated with phonotic crystal biosensors for high throughput kinetic biomolecular interaction analysis. *Lab Chip*, 7, 550-556.

[60] Goral, V., Wu, Q., Sun, H., and Fang, Y. (2011). Label-free optical biosensor with microfluidics for sensing ligand-directed functional selectivity on trafficking of thrombin receptor. *FEBS Lett.*, 585, 1054-1060.

[61] Goral, V., Jin, Y., Sun, H., Ferrie, A. M., Wu, Q., and Fang, Y. (2011). Agonist-directed desensitization of the β2-adrenergic receptor. *PLoS One,* 6, e19282.

[62] Zaytseva, N., Miller, W., Goral, V., Hepburn, J., and Fang, Y. (2011). Microfluidic resonant waveguide grating biosensor system for whole cell sensing. *Appl. Phys. Lett.*, 96, 163703.

[63] Beebe, D. J., Mensing, G. A. and Walker, G. M. (2002). Physics and applications of microfluidics in biology. *Annu. Rev. Biomed. Eng.*, 4, 261–286.

[64] Kholodenko, B. N. (2006). Cell signaling dynamics in time and space. *Nat. Rev. Mol. Cell Biol.*, 7, 165-176.

[65] Wang, M. M., Tu, E., Raymond, D. E., Yang, J. M., Zhang, H., Hagen, N., Dees, B., Mercer, E. M., Forster, A. H., Kariv, I., Marchand, P. J., and Butler, W. F. (2005). Microfluidic sorting of mammalian cells by optical force switching. *Nat. Biotechnol.*, 23, 83–87.

[66] Tourovskaia, A., Figureroa-Masot, X. and Folch, A. (2005). Differentiation-on-a-chip: a microfluidic platform for long-term cell culture studies. *Lab Chip*, 5, 14-19.

[67] Kim, L., Toh, Y.-C., Voldman, J., and Yu, H. (2007). A practical guide to microfluidic perfusion culture of adherent mammalian cells. *Lab Chip*, 7, 681-694.

[68] Lu, H., Koo, L. Y., Wang, W. M., Lauffenburger, D. A., Griffith, L. G., and Jensen, K.F. (2004). Microfluidic shear devices for quantitative analysis of cell adhesion. *Anal. Chem.*, 76, 5257-64.

[69] Giancotti, F. G. and Ruoslahti, E. (1999). Integrin signaling. *Science*, 285, 1028–1032.

[70] Engler, A. J., Sen, S., Sweeney, H. L., and Discher, D. E. (2006). Matrix elasticity directs stem cell lineage specification. *Cell*, 126, 677–689.

[71] Geiger, B., Bershadsky, A., Pankov, R., and Yamada, K. M. (2001). Transmembrane extracellular matrix–cytoskeleton crosstalk. *Nature Rev. Mol. Cell Biol.,* 2, 793-805.

[72] Folch, A. and Toner, M. (2000). Microengineering of cellular interactions. *Annu. Rev. Biomed. Eng.*, 2, 227–256.

[73] Tsang, V. L. and Bhatia, S. N. (2004). Three-dimensional tissue fabrication. *Adv. Drug Deliv. Rev.*, 56, 1635–1647.

[74] Stevens, M. M. and George, J. H. (2005). Exploring and engineering the cell surface interface. *Science*, 310, 1135–1138.

[75] Chen, C. S., Mrksich, M., Huang, S., Whitesides, G. M., and Ingber, D. E. (1997). Geometric control of cell life and death. *Science*, 276, 1425–1428.

[76] Théry, M., Racine, V., Pépin, A., Piel, M., Chen, Y., Sibarita, J. B., and Bornens, M. (2005). The extracellular matrix guides the orientation of the cell division axis. *Nat. Cell Biol.*, 7, 947–953.

[77] Sun, Y., Villa-Diaz, L. G., Lam, R. H. W., Chen, W., Krebsbach, P. H., and Fun, J. (2012). Mechanics regulates fate decisions of human embryonic stem cells. *PLoS One*, 7, e37178.

[78] Guguenguillouzo, C., Clément, B., Baffet, G., Beaumont, C., Morel-Chany, E., Glaise, D., Guillouzo, A. (1983). Maintenance and reversibility of active albumin secretion by adult-rat hepatocytes co-cultured with another liver epithelial-cell type. *Exp. Cell Res.*, 143, 47–54.

[79] Huh, D., Matthews, B. D., Mammoto, A., Montoya-Zavala, M., Hsin, H. Y., and Ingber, D. E. (2010). Reconstituting organ-level lung functions on a chip. *Science*, 328, 1662-1668.

[80] Günther, A., Yasotharan, S., Vagaon, A., Lochovsky, C., Pinto, S., Yang, J., Lau, C., Voigtlaender-Bolz, J., and Bolz, S. S. (2010). A microfluidic platform for probing small artery structure and function. *Lab Chip,* 10, 2341-2349.

[81] Domansky, K., Inman, W., Serdy, J., Dash, A., Lim, M. H., and Griffith, L.G. (2010). Perfused multiwell plate for 3D liver tissue engineering. *Lab Chip*, 10, 51-58.

[82] Sung, J. H., Kam, C. and Shuler, M. L. (2010). A microfluidic device for a pharmacokinetic-pharmacodynamic (PK-PD) model on a chip. *Lab Chip,* 10, 446-455.

[83] Baker, M. (2011). Tissue models: a living system on a chip. *Nature*, 471, 661-665.

[84] Du, F., Yu, H., Zou, B., Babcock, J., Long, S., and Li, M. (2011). hERGCentral: a large database to store, retrieve, and analyze compound-human Ether-à-go-go related gene channel interactions to facilitate cardiotoxicity assessment in drug development. *Assay Drug Dev. Technol.,* 9, 580-588.

[85] Wang, X. and Li, M. (2003). Automated electrophysiology: high throughput of art. *Assay Drug Dev. Technol.*, 1, 695-708.

[86] Wood, C., Williams, C. and Waldron, G. J. (2004). Patch clamping by numbers. *Drug Discov. Today*, 9, 434–441.

[87] Fertig, N., Blick, R. H. and Behrends, J. C. (2002). Whole cell patch clamp recording performed on a planar glass chip. *Biophys. J.*, 82, 3056–3062.

[88] Pearce, T. M., Wilson, J. A., Oakes, S. G., Chiu, S. Y., and Williams, J. C. (2005). Integrated microelectrode array and microfluidics for temperature clamp of sensory neurons in culture. *Lab Chip*, 5, 97–101.

[89] Tang, K. C., Reboud, J., Kwok, Y. L., Peng, S. L., and Yobas, L. (2010). Lateral patch-clamping in a standard 1536-well microplate format. *Lab Chip*, 10, 1044–1050.

[90] Chang, J. C., Brewer, G. J. and Wheeler, B. C. (2011). Modulation of neural network activity by patterning. *Biosens. Bioelectron.*, 16, 527–533.

[91] DeBusschere, B. D. and Kovacs, G. T. (2001). A portable cell-based biosensor system using integrated CMOS cell-cartridges. *Biosens. Bioelectron.,* 16, 543–556.

[92] Werdich, A. A., Lima, E. A., Ivanov, B., Ges, I., Anderson, M. E., Wikswo, J. P., and Baudenbacher, F. J. (2004). A microfluidic device to confine a single cardiac myocyte in a sub-nanoliter volume on planar microelectrodes for extracellular potential recordings. *Lab Chip*, 4, 357-362.

[93] Golden, A. P., Li, N., Chen, Q., Lee, T., Nevill, T., Cao, X., Johnson, J., Erdemli, G., Ionescu-Zanetti, C., Urban, L., and Holmqvist, M. (2011). IonFlux: A microfluidic patch clamp system evaluated with human *Ether-à-go-go related gene* channel physiology and pharmacology. *Assay Drug Dev. Technol.*, 9, 608-618.

[94] Huang, H., Kamm, R. D. and Lee, R. T. (2004). Cell mechanics and mechanotransduction: pathways, probes, and physiology. *Am. J. Physiol. Cell Physiol.*, 287, C1-C11.

[95] Fletcher, D. A. and Mullins, R. D. (2010). Cell mechanics and the cytoskeleton. *Nature*, 463, 485-492.

[96] Stossel, T. P., Condeelis, J., Cooley, L., Hartwig, J. H., Noegel, A., Schleicher, M., and Shapiro, S. S. (2001). Filamins as integrators of cell mechanics and signaling. *Nat. Rev. Mol. Cell Biol.*, 2, 138-145.

[97] Dos Remedios, C. G., Chhabra, D., Kekic, M., Dedova, I. V., Tsubakihara, M., Berry, D. A., and Nosworthy, N. J. (2003). Actin binding proteins: regulation of cytoskeletal microfilaments. *Physiol. Rev.*, 83, 433–473.

[98] Hou, H. W., Lee, W. C., Leong, M. C., Sonam, S., Vedula, S. R. K., and Lim, C. T. (2011). Microfluidics for applications in cell mechanics and mechanobiology. *Cell. Mol. Bioeng.*, 4, 591-602.

[99] Garcia-Cardeña, G., Comander, J., Anderson, K. R., Blackman, B. R., and Gimbrone, M. A. Jr. (2001). Biomechanical activation of vascular endothelium as a determinant of its functional phenotype. *Proc. Natl. Acad. Sci. USA,* 98, 4478-4485.

[100] Wang, J., Heo, J. and Hua, S. Z. (2010). Spatially resolved shear distribution in microfluidic chip for studying force transduction mechanism in cells. *Lab Chip*, 10, 235-239.

[101] Serra-Picamal, X., Conte, V., Vincent, R., Anon, E., Tambe, D. T., Bazellieres, E., Butler, J. P., Fredberg, J. J., and Trepat, X. (2012). Mechanical waves during tissue expansion. *Nat. Phys.*, 8, 628-634.

[102] Huang, P.-H., Lapsley, M. I., Ahmed, D., Chen, Y., Wang, L., and Huang, T. J. (2012). A single-layer, planar, optofluidic switch powered by acoustically driven, oscillating microbubbles. *Appl. Phys. Lett.,* In Press.

[103] Himmelhaus, M. and Francois, A. (2009). In-vitro sensing of biomechanical forces in live cells by a whispering gallery mode biosensor. *Biosens. Bioelectron.*, 25, 418-427.

[104] Sudhof, T. C. (2004). The synaptic vesicle cycle. *Annu. Rev. Neurosci.*, 27, 509-547.

[105] Park, J., Kile, B. M. and Wightman, R. M. (2009). *In vivo* voltammetric monitoring of norepinephrine release in the rat ventral bed nucleus of the *stria terminalis* and anteroventral thalamic nucleus. *Eur. J. Neurosci.*, 30, 2121-2133.

[106] Sun, J., Chen, P., Feng, X., Du, W., and Liu, B.-F. (2011). Development of a microfluidic cell-based biosensor integrating a millisecond chemical pulse generator. *Biosens. Bioelectron.*, 26, 3413-3419.

[107] Blagovic, K., Kim, L. Y. and Voldman, J. (2011). Microfluidic perfusion for regulating diffusible signaling in stem cells. *PLoS One*, 6, e22892.

[108] Lohse, M. J., Nikolaev, V. O., Hein, P., Hoffmann, C., Vilardaga, J. P., and Bünemann, M. (2008). Optical techniques to analyze real-time activation and signaling of G-protein-coupled receptors. *Trends Pharmacol. Sci.*, 29, 159-165.

[109] Calebiro, D., Nikolaev, V. O., Persani, L., and Lohse, M. J. (2010). Signaling by internalized G-protein-coupled receptors. *Trends Pharmacol. Sci.,* 31, 221-228.

[110] Violin, J. D., Ren, X. R. and Lefkowitz, R. J. (2006). G-protein-coupled receptor kinase specificity for β-arrestin recruitment to the β2-adrenergic receptor revealed by fluorescence resonance energy transfer. *J. Biol. Chem.,* 281, 20577-20588.

[111] Calebiro, D., Nikolaev, V. O., Gagliani, M. C., de Filippis, T., Dees, C., Tacchetti, C., Persani, L., and Lohse, M. J. (2009). Persistent cAMP-signals triggered by internalized G-protein-coupled receptors. *PLoS Biol.,* 7, e1000172.

[112] Werthmann, R. C., Volpe, S., Lohse, M. J., and Calebiro, D. (2012). Persistent cAMP signaling by internalized TSH receptors occurs in thyroid but not in HEK293 cells. *FASEB J.,* 26, 2043-2048.

[113] Swinney, D. C. (2004). Biochemical mechanisms of drug action: what does it take for success? *Nat. Rev. Drug Discov.*, 3, 801-808.

[114] Copeland, R. A., Pompliano, D. L. and Meek, T. D. (2006). Drug-target residence time and its implications for lead optimization. *Nat. Rev. Drug Discov.*, 5, 730-739.

[115] Nishijima, H., Kosaihira, A., Shibata, J., and Ona, T. (2010). Development of signaling echo method for cell-based quantitative efficacy evaluation of anti-cancer drugs in apoptosis without drug presence using high-precision surface resonance sensing. *Anal. Sci.*, 26, 529-534.

[116] Nelson, H. S. (2006). Is there a problem with inhaled long-acting β-adrenergic agonists? *J. Allergy Clin. Immunol.*, 117, 3-16.

[117] Koch, T. and Höllt, V. (2008). Role of receptor internalization in opioid tolerance and dependence. *Pharmacol. Ther.*, 117, 199-206.

[118] Rajagopal, S., Rajagopal, K. and Lefkowitz, R. J. (2010). Teaching old receptors new tricks: biasing seven-transmembrane receptors. *Nat. Rev. Drug Discov.,* 9, 373-386.

[119] Li Jeon, N., Baskaran, H., Dertinger, S. K., Whitesides, G. M., Van de Water, L., and Toner, M. (2002). Neutrophil chemotaxis in linear and complex gradients of interleukin-8 formed in a microfabricated device. *Nat. Biotechnol.*, 20, 826–830.

[120] Chung, B. G. and Choo, J. (2010). Microfluidic gradient platforms for controlling cellular behavior. *Electrophoresis*, 31, 3014-3027.

[121] Huang, Y., Agrawal, B., Sun, D., Kuo, J. S., and Williams, J. C. (2011). Microfluidics-based devices: new tools for studying cancer and cancer stem cell migration. *Biomicrofluidics*, 5, 013412.

[122] Abhyankar, V. V., Lokuta, M. A., Huttenlocher, A., and Beebe, D. J. (2006). Characterization of a membrane-based gradient generator for use in cell-signaling studies. *Lab Chip*, 6, 389–393.

[123] Pihl, J., Sinclair, J., Sahlin, E., Karlsson, M., Petterson, F., Olofsson, J., and Orwar, O. (2005). Microfluidic gradient-generating device for pharmacological profiling. *Anal. Chem.*, 77, 3897-3903.

[124] Takayama, S., Ostuni, E., LeDuc, P., Naruse, K., Ingber, D. E., and Whitesides, G. M. (2001). Laminar flows – subcellular positioning of small molecules. *Nature*, 411, 1016.

[125] Primiceri, E., Chiriaco, M. S., Dioguardi, F., Monteduro, A. G., D'Amone, E., Rinaldi, R., Giannelli, G., and Maruccio, G. (2011). Automatic transwell assay by an EIS cell chip to monitor cell migration. *Lab Chip*, 11, 4081-4086.

[126] Xu, Y., Lv, Y., Wang, L., Xing, W., and Cheng, J. (2012). A microfluidic device with passive air-bubble valves for real-time measurement of dose-dependent drug cytotoxicity through impedance sensing. *Biosens. Bioelectron.*, 32, 300-304.

[127] Grist, S., Yu, L., Chrostowski, L., and Cheung, K. C. (2012). Microfluidic cell culture systems with integrated sensors for drug screening. *Proc. SPIE*, 8251, 825103.

[128] Spiller, D. G., Wood, C. D., Rand, D. A., and White, M. R. H. (2010). Measurement of single-cell dynamics. *Nature*, 465, 736-745.

[129] Kunapuli, P., Zheng, W., Weber, M., Solly, K., Mull, R., Platchek, M., Cong, M., Zhong, Z., and Strulovici, B. (2005). Application of division arrest technology to cell-based HTS: comparison with frozen and fresh cells. *Assay Drug Dev. Technol.*, 3, 17-26.

[130] Snijder, B. and Pelkmans, L. (2011). Origins of regulated cell-to-cell variability. *Nat. Rev. Mol. Cell. Biol.*, 12, 119-125.

[131] Pelkmans, L. (2012). Using cell-to-cell variability – a new era in molecular biology. *Science*, 336, 425-426.

[132] Cohen-Saidon, C., Cohen, A. A., Sigal, A., Liron, Y., and Alon, U. (2009). Dynamics and variability of ERK2 response to EGF in individual living cells. *Mol. Cell*, 36, 885-893.

[133] Feinerman, O., Veiga, J., Dorman J. R., Germain, R. N., and Altan-Bonnet, G. (2008). Variability and robustness in T cell activation from regulated heterogeneity in protein levels. *Science*, 321, 1081-1084.

[134] Elowitz, M. B., Levine, A. J., Siggia, E. D., and Swain, P. S. (2002). Stochastic gene expression in a single cell. *Science*, 297, 1183–1186.

[135] Spencer, S. L., Gaudet, S., Albeck, J. G., Burke, J. M., and Sorger, P. K. (2009). Non-genetic origins of cell-to-cell variability in TRAIL-induced apoptosis. *Nature*, 459, 428–432.

[136] Kumaresan, P., Yang, C. J., Cronier, S. A., Blazej, R. G., and Mathies, R.A. (2008). High-throughput single copy DNA amplification and cell analysis in engineered nanoliter droplets. *Anal. Chem.*, 80, 3522–3529.

[137] Bontoux, N., Dauphinot, L., Vitalis, T., Studer, V., Chen, Y., Rossier, J., and Potier M. C. (2008). Integrating whole transcriptome assays on a lab-on-a-chip for single cell gene profiling. *Lab Chip*, 8, 443–450.

[138] Bennett, M. R. and Hasty, J. (2009). Microfluidic devices for measuring gene network dynamics in single cells. *Nature Rev. Genet.*, 10, 628–638.

[139] Bennett, M. R., Pang, W. L., Ostroff, N. A., Baumgartner, B. L., Nayak, S., Tsimring, L. S., and Hasty, J. (2008). Metabolic gene regulation in a dynamically changing environment. *Nature*, 454, 1119–1122.

[140] Locke, J. C. and Elowitz, M. B. (2009). Using movies to analyse gene circuit dynamics in single cells. *Nature Rev. Microbiol.*, 7, 383–392.

[141] Huang, B., Wu, H., Bhaya, D., Grossman A., Granier, S., Kobilka, B. K., and Zare, R. N. (2007). Counting low-copy number proteins in a single cell. *Science*, 315, 81-84.

[142] Ying, H. and Marshall, D. (2012). Microfluidics for single cell analysis. *Curr. Opin. Biotechnol.*, 23, 110-119.

[143] Zare, R. N. and Kim, S. (2010). Microfluidic platforms for single-cell analysis. *Annu. Rev. Biomed. Eng.*, 12, 187-201.

[144] Di Carlo, D., Wu, L. Y., and Lee, L. P. (2006). Dynamic single cell culture array. *Lab Chip*, 6, 1445-1449.

[145] Ghenim, L., Kaji, H., Hoshino, Y., Ishibashi, T., Haguet, V., Gidrol, X., and Nishizawa, M. (2010). Monitoring impedance changes associated with motility and mitosis of a single cell. *Lab Chip,* 10, 2546-2550.

[146] Hua, S. Z. and Pennell, T. (2009). A microfluidic chip for real-time studies of the volume of single cells. *Lab Chip*, 9, 251-256.

[147] Su, X., Qiu, Y., Marquez-Curtis, L., Gupta, M., Capjack, C. E., Rozmus, W., Janowska-Wieczorek, A., and Tsui, Y. Y. (2011). Label-free and noninvasive optical detection of the distribution of nanometer-size mitochondria in single cells. *J. Biomed. Opt.*, 2011, 16, 067003.

[148] Zheng, Y., Shojaei-Baghini, E., Azad, A., Wang, C., and Sun, Y. (2012). High-throughput biophysical measurement of human red blood cells. *Lab Chip*, 12, 2560-2567.

[149] Ang, P. K., Li, A., Jaiswal, M., Wang, Y., Hou, H. W., Thong, J. T., Lim, C. T., and Loh, K. P. (2011). Flow sensing of single cell by graphene transistor in a microfluidic channel. *Nano Lett.*, 11, 5240-5246.

[150] Kim, D., Choi, E., Choi, S. K., Lee, S., Park, J., and Yun, K.-S. (2010). Measurement of single-cell deformability using impedance analysis on microfluidic chip. *Jap. J. Appl. Phys.,* 49, 127002.

[151] Abdolahad, M., Taghinejad, M., Taghinejad, H., Janmaleki, M., and Mohajerzadeh, S. (2012). A vertically aligned carbon nanotube-based impedance sensing biosensor for rapid and high sensitive detection of cancer cells. *Lab Chip*, 12, 1183-1190.

[152] Gossett, D. R., Tse, H. T., Lee, S. A., Ying, Y., Lindgren, A. G., Yang, O. O., Rao, J., Clark, A. T., and Di Carlo, D. (2012). Hydrodynamic stretching of single cells for large population mechanical phenotyping. *Proc. Natl. Acad. Sci. USA*, 109, 7630-7635.

[153] Chen, J., Zheng, Y., Tan, Q., Shojaei-Baghini, E., Zhang, Y. L., Li, J., Prasad, P., You, L., Wu, X. Y., and Sun, Y. (2011). Classification of cell types using a microfluidic device for mechanical and electrical measurement on single cells. *Lab Chip*, 11, 3174-3181.

[154] Fang, Y. and Ferrie, A. M. (2008). Label-free optical biosensor for ligand-directed functional selectivity of acting on β_2 adrenoceptor in living cells. *FEBS Lett.*, 582, 558–564.

[155] Ferrie, A. M., Sun, H. and Fang, Y. (2011). Label-free integrative pharmacology on-target of drugs at the β_2-adrenergic receptor. *Sci. Rep.*, 1, 33.

[156] Verrier, F., An, S., Ferrie, A. M., Sun, H., Kyoung, M., Fang, Y., and Benkovic, S. J. (2011). G protein-coupled receptor signaling regulates the dynamics of a metabolic multienzyme complex. *Nat. Chem. Biol.*, 7, 909–915.

[157] Scott, C. W. and Peters, M. F. (2010). Label-free whole-cell assays: expanding the scope of GPCR screening. *Drug Discov. Today*, 15, 704-716.

[158] Mailman, R. B. (2007). GPCR functional selectivity has therapeutic impact. *Trends Pharmacol. Sci.*, 28, 390-396.

[159] Kenakin, T. and Miller, L. J. (2010). Seven transmembrane receptors as shapeshifting proteins: the impact of allosteric modulation and functional selectivity on new drug discovery. *Pharmacol. Rev.*, 62, 265-304.

[160] Han, Y., Moreira, I. S., Urizar, E., Weinstein, H., and Javitch, J. A. (2009). Allosteric communication between protomers of dopamine class A GPCR dimers modulates activation. *Nat. Chem. Biol.*, 5, 688-695.

[161] Stallaert, W., Dorn, J. F., van der Westhuizen, E., Audet, M., and Bouvier, M. (2011). Impedance responses reveal β_2-adrenergic receptor signaling pluridimensionality and allow classification of ligands with distinct signaling profiles. *PLoS One*, 7, e29420.

In: Microfluidics
Editors: Stefano Panzarella and Walter Maroni

ISBN: 978-1-62417-345-5

Chapter 3

CONTROLLABLE MICROFLUIDIC GENERATION OF MONODISPERSE MULTIPLE EMULSION DROPLETS

Wei Wang and Liang-Yin Chu*
School of Chemical Engineering, Sichuan University,
Chengdu, Sichuan, China

ABSTRACT

Multiple emulsions, or "emulsions of emulsions", are complex nested liquid systems, in which dispersed droplets contain smaller droplets inside. They are widely used to encapsulate active ingredients in myriad applications, including drug delivery systems, foods, cosmetics, chemical separations, and syntheses of microspheres and microcapsules. Accurate control of size monodispersity and internal structure are critical for the versatility of such emulsions, because these attributes allow precise manipulation of the encapsulation of substances and the structure of synthesized materials. Compared with other emulsification techniques, such as multistep bulk emulsification and membrane emulsification, recently developed microfluidic technique shows much greater power in controlled generation of monodisperse multiple emulsion droplets. In this Chapter, droplet microfluidic device with highly scalability for generating monodisperse multiple emulsion droplets are firstly introduced, from which the readers can see the precise control on both the size and number of the inner droplets in multiple emulsions; then, based on this device, novel devices for controllable encapsulating droplets with different components in each level of multiple emulsions are introduced; the size, number, and ratio of the multicomponent inner droplets are under excellent control, which largely extend the emulsion structures and benefit the application of multiple emulsions.

Finally, the application of these multiple emulsions are demonstrated by template synthesis of functional microcapsules with highly monodispersity, controlled inner structures and encapsulation characteristics.

* Email: chuly@scu.edu.cn.

Keywords: Microfluidics; Multiple emulsions; Multicomponent; Template synthesis; Poly(*N*-isopropylacrylamide)

1. Introduction

Multiple emulsions, or "emulsions of emulsions", are complex systems in which dispersed droplets themselves contain smaller droplets inside. They are widely used to encapsulate active ingredients in myriad applications, including drug delivery systems [1-3], foods [4-5], cosmetics [6-7], chemical separations [8-9], and syntheses of microspheres and microcapsules [10-16]. Precise control of the size monodispersity and internal structure of multiple emulsions are critical for the versatility of such emulsions, because these attributes allow precise manipulation of the loading levels and the release and transport kinetics of the encapsulated substances [10-21], and the internal structures of the materials that are templated from these emulsions [10, 22]. Multiple emulsions are typically formed through sequential bulk emulsification using shear. This usually results in polydisperse emulsions. Controlling the shear with constrained geometry [23-27], porous membranes [17-18], or microchannels [19] can produce emulsions that are nearly monodisperse; these can then themselves be emulsified to create multiple emulsions. However, although the volume fractions of both the initial and final emulsions can be controlled [17-19, 23, 25-27], precise control over the number of inner droplets as well as the internal structure of the emulsions is difficult with these techniques. Microfluidic devices offer an alternate route to produce monodisperse multiple emulsions. Cascading two T-junctions [20, 28] can produce monodisperse double emulsions with some control over the number and size of the inner droplets; however, the standard two-dimensional microfluidic channels require precise, localized modification of the surface chemistry to control the wettability to enable the production of multiple emulsions; this ultimately limits their utility. By contrast, coaxial flow-focusing [10, 12-13] geometries relax the wetting constraints, but are still limited in the range of fluids that can be used and in the precise control afforded over the number and size of inner droplets. In addition, it is still difficult for these microfluidic methods to scale up for the generation of higher order multiple emulsions, such as triple emulsions. Recently, a novel type of scalable microfluidic devices that based on assembly of glass capillaries have been developed for generating monodisperse multiple emulsions with precise control of both the size and number of the inner droplets; scale-up of the devices allows the controllable production of higher order multiple emulsions [29]. Further improvement of this capillary microfluidic technique creates novel devices for controllable encapsulating droplets with different components in each level of multiple emulsions, where the size, number, and ratio of the multicomponent inner droplets are under excellent control [30]. These techniques largely extend the emulsion structures and benefit the application of multiple emulsions. In this chapter, hierarchical and scalable microfluidic devices made of glass capillaries for generating monodisperse multiple emulsions with precisely controlled structures will be introduced, and the prospect of developments and applications of these multiple emulsions will be discussed.

2. Controllable Production of Monodisperse Multiple Emulsions

Monodispersity of size and accurate control of internal structure are critical for the versatility of such emulsions because these attributes allow precise manipulation of the encapsulation of substances [10-21] and the structure of synthesized materials [10, 22]. For example, with accurate control of the number and size of inner droplets, transport kinetics of encapsulated substances can be precisely manipulated.

Control of the number of inner drops is also critical for structuring colloidal assemblies to produce non-spherical particles [31]. Furthermore, if the multiple emulsions are used as templates for synthesis of novel materials, with the inner droplets serving as separate compartments for encapsulation, precise control of the internal structure of emulsion templates is again essential.

In this section, we fabricate highly monodisperse multiple emulsions using capillary microfluidics to independently control both the size and the number of inner droplets. Importantly, our technique is easily scalable to higher order multiple emulsions, which we demonstrate by making monodisperse, controllable triple emulsions.

2.1. Controllable Monodisperse Double Emulsions

2.1.1. Fabrication of the Capillary Microfluidic Device

We present a highly scalable micro-capillary technique that simultaneously controls the droplet monodispersity as well as the number and size of the inner droplets. The fabrication process of double emulsions by this micro-capillary technique includes two emulsification steps. The first emulsification step is accomplished using a coaxial, co-flow geometry. It is comprised of the injection tube, a cylindrical capillary with a tapered end, which is inserted into the transition tube, a second cylindrical capillary with an inner diameter, D_2, as shown in Figure 1.

The outer diameter of cylindrical capillaries was 1 mm. The square capillary tubes had an inner dimension of 1 mm. Both cylindrical capillary tubes are centered within the larger square capillary with alignment ensured by matching the outer diameters of the cylindrical tubes to the inner dimension of the square one. The other end of the transition tube is again tapered and inserted into a third, coaxially aligned cylindrical capillary tube, the collection tube, of inner diameter, D_3. For the double emulsions, the inner diameters (ID) of the transition tube and the collection tube were 200 μm and 580 μm, respectively. A transparent epoxy resin was used to seal the tubes where required.

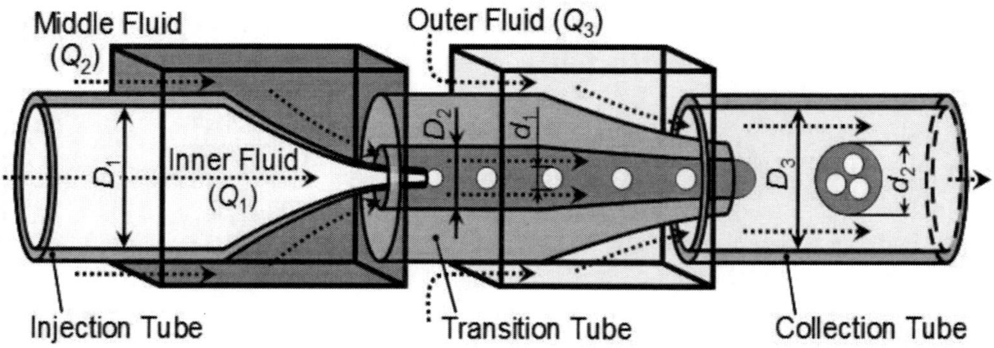

(Reproduced with permission from Ref. [29]). Copyright (2007), Wiley-VCH Verlag GmbH and Co. KGaA.

Figure 1. Schematic diagram of the capillary microfluidic device. The outer fluid must be immiscible with the middle fluid and the middle fluid must be immiscible with the inner fluid.

2.1.2. Generation of Monodisperse Double Emulsions

To generate emulsion droplets from the capillary microfluidic device, solutions were supplied to the device through polyethylene tubing attached to syringes. The outer fluid for the double emulsions was poly(dimethylsiloxane) oil (PDMS) with a viscosity of 100 cSt, containing 2 wt% Dow Corning 749 fluid as a surfactant. The middle fluid was an aqueous solution containing 10 wt% glycerol and 2 % (w/v) poly(vinyl alcohol) (PVA, 87-89 % hydrolyzed). The inner fluid was PDMS oil with a viscosity of 10 cSt. The fluid flow was driven with positive displacement syringe pumps. The drop formation was imaged with a high-speed camera attached to a microscope.

Uniform monodisperse multiple emulsions are generated in two emulsification steps: Droplets of the inner fluid are emulsified in the first stage of the device by coaxial flow of the middle fluid (Figure 2a).

The single emulsion is subsequently emulsified in the second stage through coaxial flow of the outer fluid, which is injected in the outer stream through the square capillary (Figure 1). In both emulsification steps, droplets form immediately at the exit of the tapered capillary (Figures 2a and 2b); this dripping mechanism [32] ensures formation of highly monodisperse drops. The separation of the two emulsification steps affords independent control over each; this control was achieved by adjusting the device dimensions as well as the inner, middle and outer fluid flow rates, Q_1, Q_2, and Q_3, respectively. The number of inner droplets can be precisely controlled, as illustrated by the double emulsions containing one through eight inner droplets, shown in Figure 2c. Similarly, the size of the inner droplets can be precisely controlled, as shown by the three sets of double emulsions with different inner droplet sizes in Figure 2d.

In each case, all droplets formed are highly monodisperse. The coefficient of variation (*CV*, defined as the ratio of the standard deviation of the size distribution to its arithmetic mean) is used to characterize the size monodispersity of emulsion droplets. In all the experiments, the *CV* values for diameters of internal droplets and for double emulsions are always less than 2.3 % and 1.6 %, respectively. In addition, the coaxial structure of this capillary microfluidic device has the advantage that no surface modification of wettability is necessary, allowing the same device to be used to prepare either water-in-oil-in-water (W/O/W) or the inverse, oil-in-water-in-oil (O/W/O) multiple emulsions.

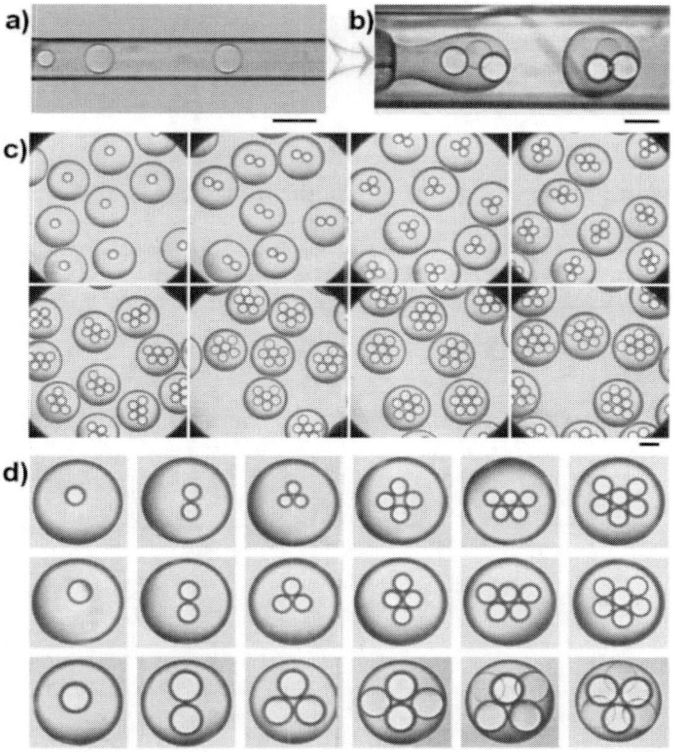

(Reproduced with permission from Ref. [29]). Copyright (2007), Wiley-VCH Verlag GmbH and Co. KGaA.

Figure 2. Formation of the precisely controlled monodisperse double emulsions. (a,b) High-speed optical micrographs of the (a) first and (b) second emulsification stages. (c) Optical micrographs of monodisperse double emulsions containing a controlled number of monodisperse single emulsions. (d) Optical micrographs of monodisperse double emulsions showing controlled increase of the diameter of the inner droplets while the number is constant. All double emulsions were made in the same device and with the same fluids. The flow rates of the inner, middle and outer fluids in (a) and (b) are $Q_1 = 350$ μL·h^{-1}, $Q_2 = 2000$ μL·h^{-1}, and $Q_3 = 5000$ μL·h^{-1}, respectively. In (c), the flow rates of middle and outer fluids are fixed as $Q_2 = 2000$ μL·h^{-1} and $Q_3 = 5000$ μL·h^{-1}, and those of the inner fluid are $Q_1 = 20$, 55, 70, 85, 150, 200, 225, and 240 μL·h^{-1}, from the left top to the right bottom. In (d), the variation ranges for the flow rates of the inner, middle and outer fluids are $Q_1 = 20$~600 μL·h^{-1}, $Q_2 = 1600$~5000 μL·h^{-1}, and $Q_3 = 2000$~8000 μL·h^{-1} (in each case Q_3 is always larger than Q_2). All scale bars are 200 μm.

2.1.3. Prediction of the Number of Inner Droplets

The high degree of control and stability afforded by the dripping mechanism allows us to quantitatively determine the flow-rate dependence of the diameters of the inner and outer drops, d_1 and d_2, respectively, and to predict the number of inner droplets, N_1. For fixed device dimensions and solution conditions, the droplet diameter in coaxial flow is inversely proportional to the velocity of the surrounding flow [32], in the dripping regime. We calibrate the dependence of d_1/D_2 on Q_1/Q_2, and find the linear dependence depicted in Figure 3a. We find a similar linear dependence of d_2/D_3 on $(Q_1 + Q_2)/Q_3$, shown in Figure 3b. Using these empirical relations, we can quantitatively predict the number of inner droplets, $N_1 = f_1 / f_2$, where f_1 and f_2 are the formation rates of the inner and outer droplets. From mass conservation for each stream,

$$N_1 = \frac{f_1}{f_2} = \frac{Q_1/(\pi d_1^3/6)}{(Q_1+Q_2)/(\pi d_2^3/6)} = \frac{Q_1}{Q_1+Q_2}\cdot\frac{d_2^3}{d_1^3} \quad (1)$$

Using the linear fits to the experimental data in Figures 3a and 3b, we can predict the number of encapsulated droplets as a function of the flow rates,

$$N_1 = \frac{Q_1}{Q_1+Q_2}\cdot\frac{D_3^3}{D_2^3}\cdot\left(\frac{a_2(Q_1+Q_2)/Q_3+b_2}{a_1(Q_1/Q_2)+b_1}\right)^3 \quad (2)$$

where, a_1 and a_2 are the slopes, and b_1 and b_2 are the intercepts obtained from the fits in Figures 3a and 3b, respectively.

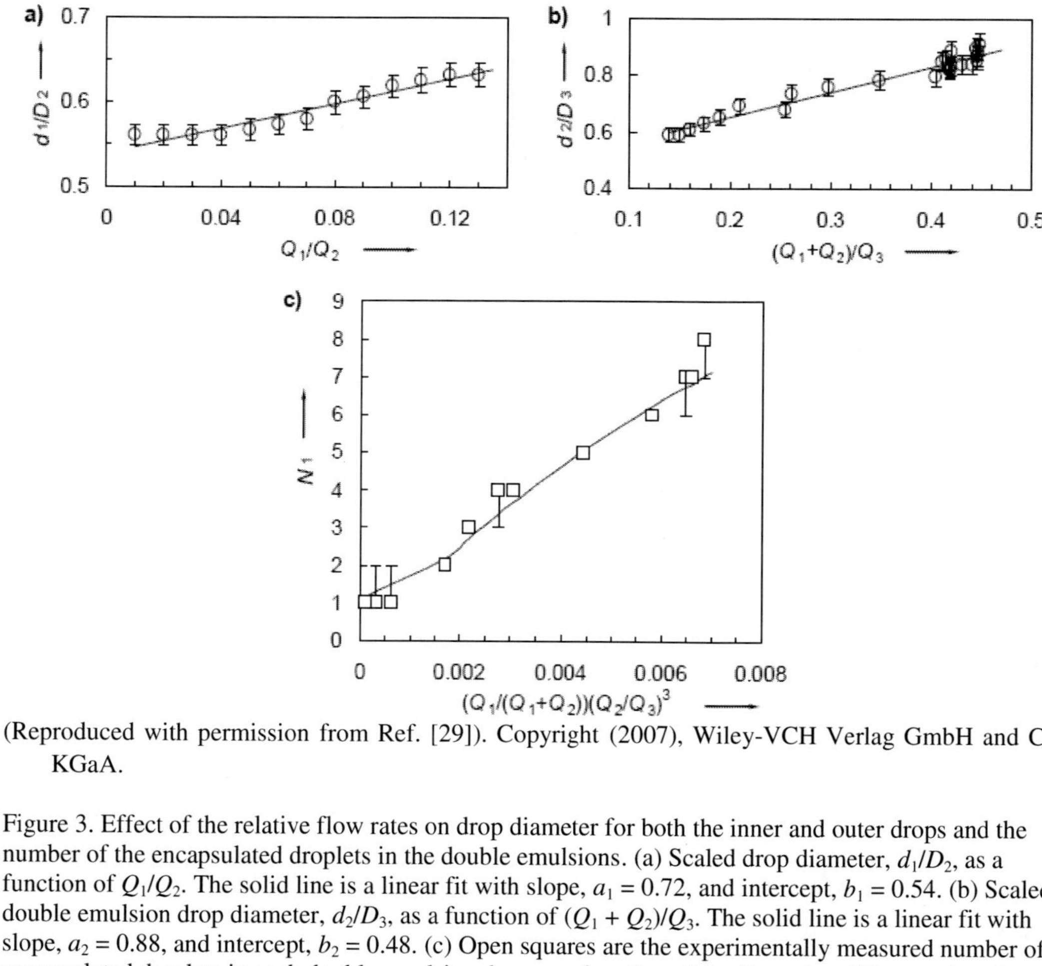

Figure 3. Effect of the relative flow rates on drop diameter for both the inner and outer drops and the number of the encapsulated droplets in the double emulsions. (a) Scaled drop diameter, d_1/D_2, as a function of Q_1/Q_2. The solid line is a linear fit with slope, $a_1 = 0.72$, and intercept, $b_1 = 0.54$. (b) Scaled double emulsion drop diameter, d_2/D_3, as a function of $(Q_1 + Q_2)/Q_3$. The solid line is a linear fit with slope, $a_2 = 0.88$, and intercept, $b_2 = 0.48$. (c) Open squares are the experimentally measured number of encapsulated droplets in each double emulsion drop as a function of the flow rates. We note that d_1 is inversely proportional to Q_2 and d_2 is inversely proportional to Q_3; thus rather than using the expression in Eqation (1), we plot the data as a function solely of flow rates. Equation (2) is used to predict the number of encapsulated drops, depicted by the line. We use the slopes and intercepts from (a) and (b) in Eqation (2) in this calculation. The variation ranges for the flow rates of the inner, middle and outer fluids are 20~260 $\mu L\cdot h^{-1}$, 2000 $\mu L\cdot h^{-1}$, and 5000~15000 $\mu L\cdot h^{-1}$.

The experimentally measured N_1 are well described by Equation (2), with no free parameters, as shown by the solid line in Figure 3c; this allows easy calibration of our device. When the number predicted is an integer, N_1 can be precisely controlled; however, when the number predicted between two integers, the operation is in a transition zone, and N_1 takes either integral value. However, precise control over N_1 can be achieved through correct adjustment of the flow rates.

2.2. Controllable Monodisperse Triple Emulsions

This method for fabricating multiple emulsions has the significant advantage that it can be very easily extended, enabling us to generate additional hierarchical levels of multiple emulsions. We illustrate this by fabricating monodisperse triple emulsions, which consist of water-in-oil-in-water-in-oil (W/O/W/O) drops.

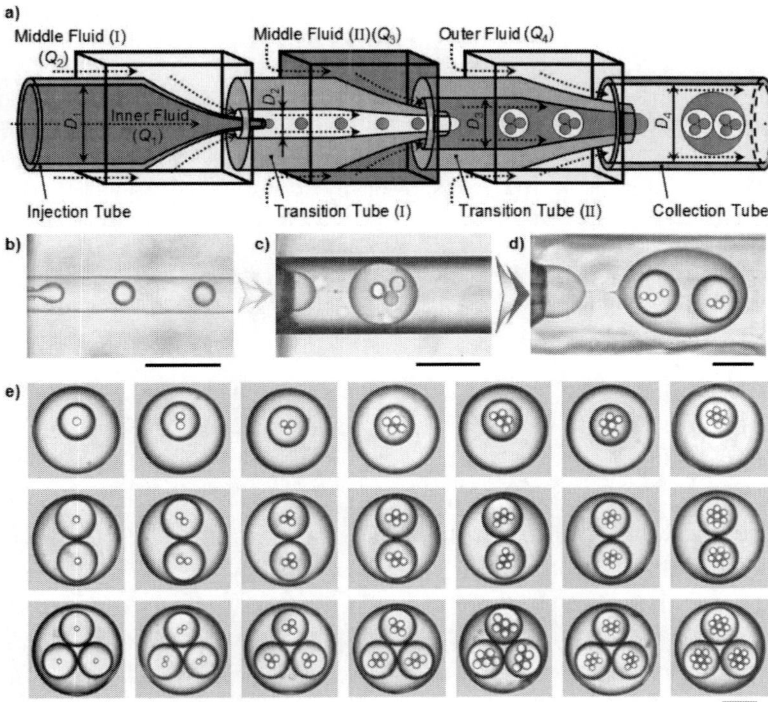

Figure 4. Generating highly controlled monodisperse triple emulsions. (a) Schematic diagram of the extended capillary microfluidic device for generating triple emulsions. (b-d) High-speed optical micrographs displaying the (b) first, (c) second, and (d) third emulsification stages. The flow rates of the inner, middle (I), middle (II) and outer fluids in (b-d) are Q_1 = 50 μL·h^{-1}, Q_2 = 500 μL·h^{-1}, Q_3 = 2000 μL·h^{-1}, and Q_4 = 5000 μL·h^{-1}. (e) Optical micrographs of triple emulsions that contain a controlled number of inner and middle droplets. The variation ranges for the flow rates in (e) are Q_1 = 5~100 μL·h^{-1}, Q_2 = 200~1000 μL·h^{-1}, Q_3 = 2000~3500 μL·h^{-1}, and Q_4 = 5000 μL·h^{-1}. The scale bar in all images is 200 μm.

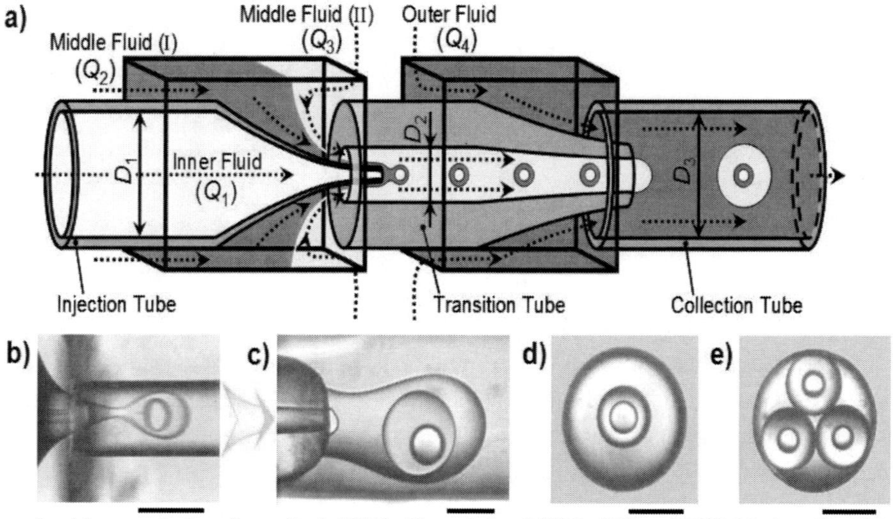

(Reproduced with permission from Ref. [29]). Copyright (2007), Wiley-VCH Verlag GmbH and Co. KGaA.

Figure 5. Generating highly controlled monodisperse triple emulsions. (a) Schematic diagram detailing an alternate method for generating triple emulsions where the middle fluid (II) is injected from the entry side of the first square tube, leading to flow-focusing of the first middle fluid into the transition capillary. (b,c) High-speed optical micrographs showing the formation of (b) double emulsions in a one-step process in the transition capillary and the subsequent formation of (c) triple emulsions in the collection capillary. (d,e) Optical micrographs of triple emulsions that contain a different number of double emulsions. The variation ranges for the flow rates in (b-e) are Q_1 = 50~200 μL·h^{-1}, Q_2 = 1600~2500 μL·h^{-1}, Q_3 = 4000~8000 μL·h^{-1}, and Q_4 = 5000~12000 μL·h^{-1}. The scale bar in all images is 200 μm.

This is accomplished by adding a second transition tube at the outlet of the first, and injecting the outermost fluid to flow coaxially around this tube to form the third level of emulsification at its outlet, as shown in Figure 4a. In this microfluidic device, the IDs of the first (I) and second (II) transition capillaries were 100 μm and 250 μm, respectively. The ID of the collection tube was 580 μm. For the triple emulsions, the outer fluid was PDMS oil with a viscosity of 100 cSt, containing 2 wt% Dow Corning 749 fluid as a surfactant. The middle fluid (II) was an aqueous solution containing 10 wt% glycerol and 2 % (w/v) PVA (87-89 % hydrolyzed). The middle fluid (I) was kerosene containing 10 % (w/v) polyglycerol polyricinoleate (PGPR 90) and 15 wt% Dow Corning 749 fluid. The inner fluid was the same aqueous solution containing 10 wt% glycerol and 2 % (w/v) PVA. The individual steps of drop formation leading to the triple emulsions are shown in Figure 4b-d. Although there are large deformations of the droplets during their formation as they flow through the tapered regions of the capillaries, the emulsification process remains stable. Again, both the diameter and number of the individual drops, at every level, are precisely controllable, as illustrated by the series of drops with the innermost varying in number from one to seven, and middle varying in number from one to three, shown in Figure 4e.

In all experiments, the *CV* of the diameters of every batch of triple emulsions is always less than 1.5 %. Furthermore, the technique can clearly be sequentially scaled to even higher levels of emulsification if desired; for example, quadruple emulsions could be made by adding an additional stage.

The simplicity of this method also enables us to incorporate alternate emulsification schemes. For example, we can inject a second fluid at the inlet of the transition tube to create double emulsions by flow-focusing of a coaxial stream [10], followed by an additional emulsification step at the outlet of the transition tube to create triple emulsions, as shown in Figure 5a. While only two drop formation steps are required (Figures 5b and 5c), the ability to independently control the number and size of innermost drops is reduced. Nevertheless, highly monodisperse triple emulsions can be fabricated, albeit with only a single innermost droplet (Figures 5d and 5e).

3. Controllable Production of Monodisperse Multicomponent Multiple Emulsions

Multiple emulsions containing droplets with different contents in the same internal level are important for co-encapsulating incompatible actives or chemicals for synergistic delivery or for biochemical or chemical reactions, and for fabricating hierarchical multicompartment materials [22, 33].

Microfluidic techniques are an excellent method for manipulation of emulsions [10, 29, 34-39]; both glass capillary devices [10, 29] and poly(dimethylsiloxane) devices [34-37] enable precise preparation of monodisperse multiple emulsions, with the size and number of droplets at each level in multiply nested emulsions being precisely controlled. However, control of the number, ratio and size of different droplets within each level, and the co-encapsulation of multicomponent droplets for higher-order multiple emulsions, can not be easily achieved with these devices.

Controlled, higher-order, multicomponent multiple emulsions would offer advanced platforms for design of more complex multicompartment materials, and would provide synergistic delivery systems or chemical microreactors for incompatible actives or chemicals with more precise encapsulation and regulated mass-transfer kinetics. Co-encapsulation of two droplets containing different contents in double emulsions has been accomplished with glass microfluidic devices using a cross-junction structure [20, 28] and with capillary microfluidic devices with two separate internal channels in the injection tube [33]. However, precise control over the number, ratio and size of the co-encapsulated multicomponent droplets in double emulsions remains very difficult to achieve, and this is essential to fully exploit the power and potential of these structures.

Thus, new techniques for controlled generation of higher-order multicomponent multiple emulsions are required.

In this section, we will introduce a hierarchical and scalable microfluidic device that generates multicomponent multiple emulsions where the number, ratio and size of the co-encapsulated droplets, containing different contents, are precisely controlled at each level. We demonstrate the scalability and controllability of the microfluidic device by starting with quadruple-component double emulsions, and scaling up by simply adding building blocks to generate quintuple-component double emulsions, quintuple-component triple emulsions, and even sextuple-component triple emulsions. In each case, the number, ratio and size of co-encapsulated multicomponent droplets are individually controlled with unprecedented accuracy.

3.1. Controllable Production of Monodisperse Quadruple-Component Double Emulsions

3.1.1. Design and Fabrication of Microfluidic Device for Multicomponent Multiple Emulsions

The novel microfluidic device for multicomponent multiple emulsions is comprised of three basic units, a drop maker, a connector and a liquid extractor (Figure 6). The drop maker is designed for generating droplets (Figure 6a); the connector is designed for merging droplets from different drop makers (Figure 6b), and the liquid extractor is designed for removing unwanted fluid from the continuous phase (Figure 6c). By using different combinations of the building blocks we achieve a versatile route to creating microfluidic devices for the formation of multicomponent multiple emulsions with highly controlled yet exceptionally diverse structures.

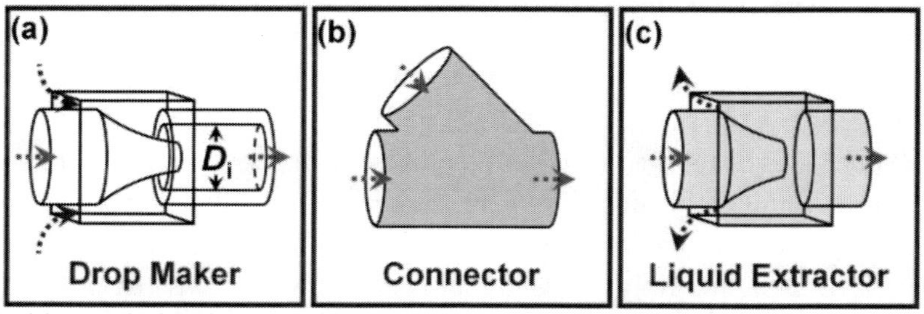

Figure 6. Schematic diagram of functional building blocks for microfluidic device. (a) Drop maker for generating droplets, (b) connector for merging droplets from different drop makers, and (c) liquid extractor for removing excess continuous phase.

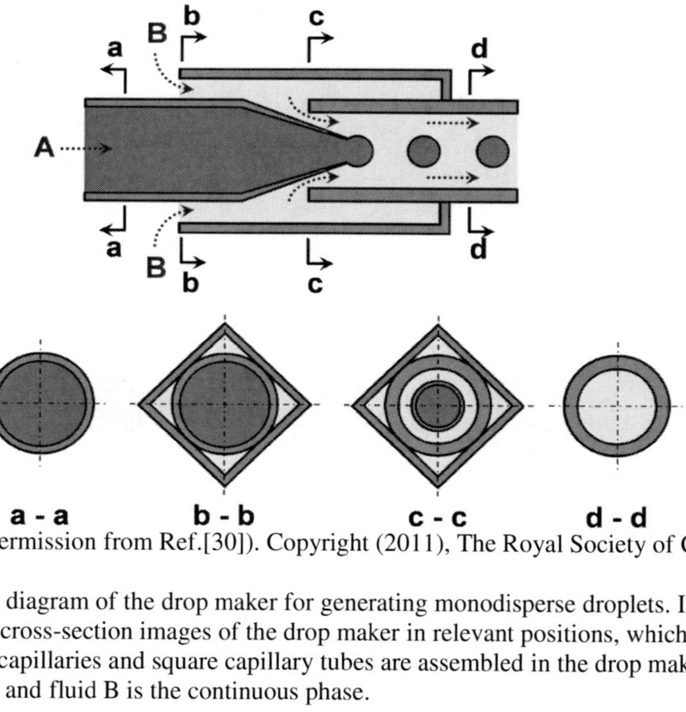

Figure 7. Schematic diagram of the drop maker for generating monodisperse droplets. Illustrations a-a, b-b, c-c and d-d are cross-section images of the drop maker in relevant positions, which clearly show how the cylindrical capillaries and square capillary tubes are assembled in the drop maker. Fluid A is the dispersed phase, and fluid B is the continuous phase.

Cylindrical capillaries with outer diameter of 1.0 mm are used as the injection tube and collection tube. The end of the injection tube is tapered by a micropuller and then adjusted by a microforge. Square capillary tubes with inner dimension of 1.0 mm are used to ensure the coaxiality of the cylindrical capillaries nested within them. For the drop maker, the injection tube and collection tube are aligned inside the square capillary tube, with the tapered end of the injection tube inserted into the collection tube. The coaxial co-flow geometry of drop maker is ensured by matching the outer diameters of the cylindrical capillaries to the inner dimensions of the square capillary tubes, as shown in Figure 7. The dispersed fluid A is pumped into the injection tube, and the continuous fluid B is pumped through the outer coaxial region between the square capillary tube and the cylindrical injection tube. The coaxial flow of these two fluids results in the formation of monodisperse droplets in the collection tube. For the connector, two cylindrical capillaries are integrated to build a Y-shaped connector. For the liquid extractor, the injection tube and collection tube are aligned inside the square capillary tube but with the tapered end of the injection tube out of the collection tube. Fabrication of the microfluidic devices for various multicomponent multiple emulsions is achieved by selective bonding of the drop maker, connector and liquid extractor.

3.1.2. Controllable Generation of Monodisperse Quadruple-Component Double Emulsions

We demonstrate the versatility and controllability of microfludic devices based on these building blocks by first generating quadruple-component double emulsions, in each of which droplets with different contents can be co-encapsulated while precisely and independently controlling the number, ratio and size of the droplets.

The multicomponent multiple emulsions are generated by operating fluids of aqueous phase and oil phase into and out of the microfluidic device through syringe pumps. With 5 % (w/v) polyglycerol polyricinoleate (PGPR 90) as surfactant, soybean oil and benzyl benzoate are used as two kinds of oil phases. Aqueous solution containing 1 % (w/v) Pluronic F127 and 5 % (w/v) glycerol is used as aqueous phase. Either 0.1 % (w/v) Sudan red or 0.1 % (w/v) Sudan black is used to color the oil phase, and 0.5 % (w/v) methylene blue is used to color the aqueous phase. For quadruple-component double emulsions, inner fluids $F_{1\text{-}1}$ and $F_{1\text{-}2}$ are soybean oil phases with and without Sudan red respectively, $F_{2\text{-}1}$, $F_{2\text{-}2}$, $F_{2\text{-}C}$ and $F_{2\text{-}E}$ are aqueous phases, and $F_{3\text{-}M}$ is soybean oil phase, as shown in Figure 8. The variation ranges for the flow rates are $Q_{1\text{-}1} \approx 50\text{~}200$ μL·h^{-1}, $Q_{1\text{-}2} \approx 60\text{~}300$ μL·h^{-1}, $Q_{2\text{-}1} \approx 200\text{~}600$ μL·h^{-1}, $Q_{2\text{-}2} \approx 300\text{~}700$ μL·h^{-1}, $Q_{2\text{-}C} \approx 150\text{~}1600$ μL·h^{-1}, $Q_{2\text{-}E} \approx 100\text{~}700$ μL·h^{-1} and $Q_{3\text{-}M} \approx 4000\text{~}11000$ μL·h^{-1}.

As shown in Figure 8, the microfluidic device comprises three drop makers, two connectors and one liquid extractor. First, droplets of two different inner fluids ($F_{1\text{-}1}$ and $F_{1\text{-}2}$) are separately emulsified in drop makers in different branch channels. Then, through the connectors, these two sets of droplets flow into the main channel, where they are merged to form a linear array of droplets.

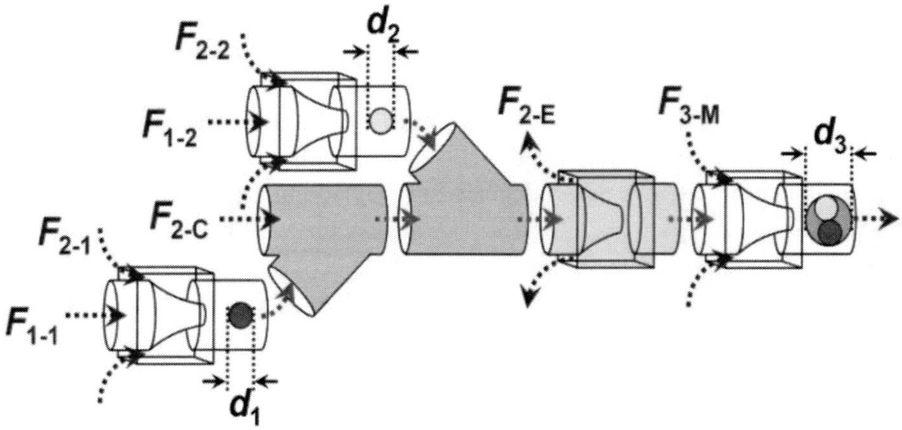

Figure 8. Schematic diagram of microfluidic device constructed from the building blocks for generating quadruple-component double emulsions, in which d_1, d_2 and d_3 are the diameters for inner red, inner transparent, and outer droplets respectively. $F_{i\text{-}j}$ is fluid phase for each level of multiple emulsions, where the subscript *i* refers to the level number where $F_{i\text{-}j}$ exists in multiple emulsions by counting from the inside outward, and the subscript *j* refers to the injection position of $F_{i\text{-}j}$ in the microfluidic device. The subscript *j* can be number or letter, where number (1, 2, ...) means the number of branch channel, and letter *C*, *E* and *M* stand for connector, liquid extractor and main channel respectively.

Because of the periodic droplet formation in each drop maker, the two types of droplets are regularly and alternately aligned in the main channel. We control the spacing between the droplets by using a connector ($F_{2\text{-}C}$) to inject fluid and a separate extractor ($F_{2\text{-}E}$) to remove the middle fluid as required. The middle fluid is further emulsified by outermost fluid in the drop maker ($F_{3\text{-}M}$) downstream, forming double emulsions containing the two different types of droplets.

For quadruple-component double emulsions, the inner red and transparent droplets are separately formed from the drop makers in different branch channels, so the formation rate of each kind of droplets can be individually tuned by changing the relevant flow rates, which results in a regular array of red and transparent droplets with controlled ratio in the main channel, as shown in Figure 9. When the formation rates of two kinds of droplets are equal, the red and transparent droplets arrange in the main channel one by one, where one red droplet and one transparent droplet makes a periodic unit of the regular droplet array (Figure 9a). Break-up in the intervals of periodic units leads to co-encapsulation of two kinds of droplets with controlled ratio at 1:1. By changing the flow rate of the outer fluid, every one, two and three periodic units can be encapsulated in double emulsions, resulted in quadruple-component double emulsions containing two, four and six droplets respectively, with the ratio (R_{12}) of the red droplets to transparent droplets fixed at 1:1 (Figure 9a). By increasing the formation rate of the red droplets, regular arrays with more red droplets in every periodic unit can be obtained (Figures 9b-d). Break-up in the intervals of these periodic units results in quadruple-component double emulsions with R_{12} controlled at 2:1, 3:1 and 3:2, as shown in Figures 9b, 9c, and 9d, respectively. Similarly, increase in the formation rate of the transparent droplets leads to regular arrays with more transparent droplets in every periodic unit (Figures 9e-g), and periodic break-up of these arrays results in quadruple-component double emulsions with R_{12} controlled at 1:2, 1:3 and 2:3, as shown in Figures 9e, 9f, and 9g, respectively. Therefore, the number and ratio of the encapsulated droplets with different contents can be independently and precisely controlled by adjusting the flow rates of relevant fluids.

The controlled production of monodisperse quadruple-component double emulsions, made from oil-in-water-in-oil (O/W/O), is shown in Figure 10. The number and ratio of inner red and transparent droplets are precisely manipulated. Accurate regulation of number and ratio is achieved by fine adjustment of the flow rates. We demonstrate the excellent controllability of our microfluidic device by independently changing the number of encapsulated inner red (N_r) and transparent (N_t) droplets; this also leads to precise control over the ratio (R_{12}) of red droplets to transparent droplets (Figure 10). We illustrate this control by showing the resultant double emulsions as a plot of N_r as a function of N_t, with each varying from 1 to 3, in Figure 10.

Independent control over N_r and N_t affords accurate manipulation of R_{12} with desired values. Because we use the dripping mechanism to provide highly control over the formation of droplets in our microfluidic device, the droplets are generated with good reproducibility and monodispersity.

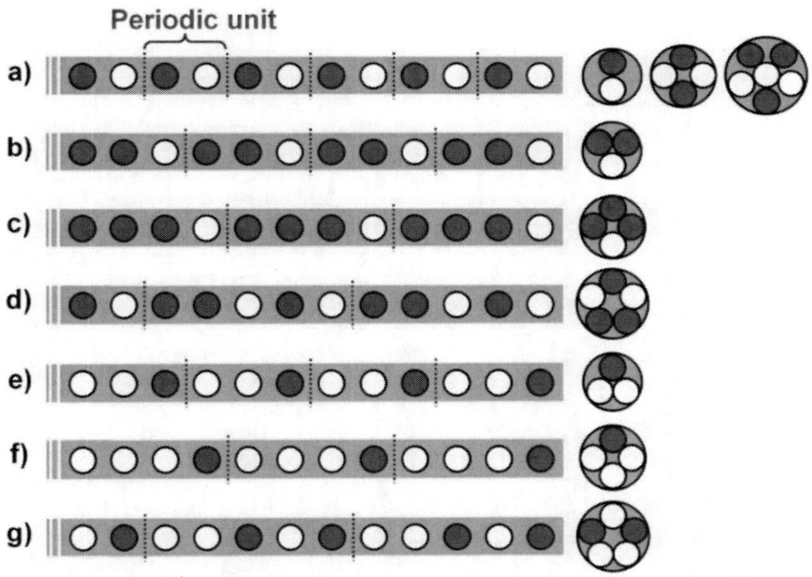

Figure 9. Schematic diagram of controlled generation of quadruple-component double emulsions. Break-up of these regular droplet arrays in the intervals of periodic units results in co-encapsulation of two different droplets with precisely controlled number and ratio.

We can also achieve excellent control over droplet size (d_i) by changing the flow rates or the inner diameter (D_i) of the collection tube of drop makers. The size manipulation is illustrated by controllably generating quadruple-component double emulsions that contain smaller red droplets and larger blue droplets (Figure 11). Here we use drop makers with different D_i in branch channels to generate the smaller red droplets and larger blue droplets (Figure 8). Also, the number and ratio of these two different droplets can be controlled by adjustment of the flow rates. The number of smaller red droplets increases from 0 to 5 while that of larger blue droplets keeps fixed at 1, resulted in a controlled ratio of smaller red droplets to larger blue droplets changing from 1:1 to 5:1, as shown in Figure 11b-f. For generation of quadruple-component double emulsions containing smaller red droplets and larger blue droplets, the inner fluids $F_{1\text{-}1}$ and $F_{1\text{-}2}$ are benzyl benzoate oil phases with Sudan red and Sudan black respectively, $F_{2\text{-}1}$, $F_{2\text{-}2}$, $F_{2\text{-}C}$ and $F_{2\text{-}E}$ are aqueous phases, and $F_{3\text{-}M}$ is soybean oil phase, as shown in Figure 8. The variation ranges for the flow rates are $Q_{1\text{-}1} \approx$ 0~350 μL·h^{-1}, $Q_{1\text{-}2} \approx$ 250~750 μL·h^{-1}, $Q_{2\text{-}1}$ =800 μL·h^{-1}, $Q_{2\text{-}2}$ = 2000 μL·h^{-1}, $Q_{2\text{-}C} \approx$ 1500~1800 μL·h^{-1}, $Q_{2\text{-}E}$ = 200 μL·h^{-1} and $Q_{3\text{-}M} \approx$ 4000~5000 μL·h^{-1}.

These results demonstrate successful co-encapsulation of droplets with different contents in double emulsions, with the number, ratio and size of different inner droplets precisely and independently controlled. These co-encapsulated different droplets can be employed to separately load incompatible actives or chemicals, and can be employed as different compartments for multicompartment materials. The excellent control over number, ratio and size of different inner droplets allows us to precisely and independently manipulate the encapsulation of each active or chemical and the internal structure of multicompartment materials. Furthermore, generation of quadruple-component double emulsions with inverse water-in-oil-in-water (W/O/W) type is also feasible with our device because its coaxial structure does not require modification of surface wettability for drop formation of different fluids.

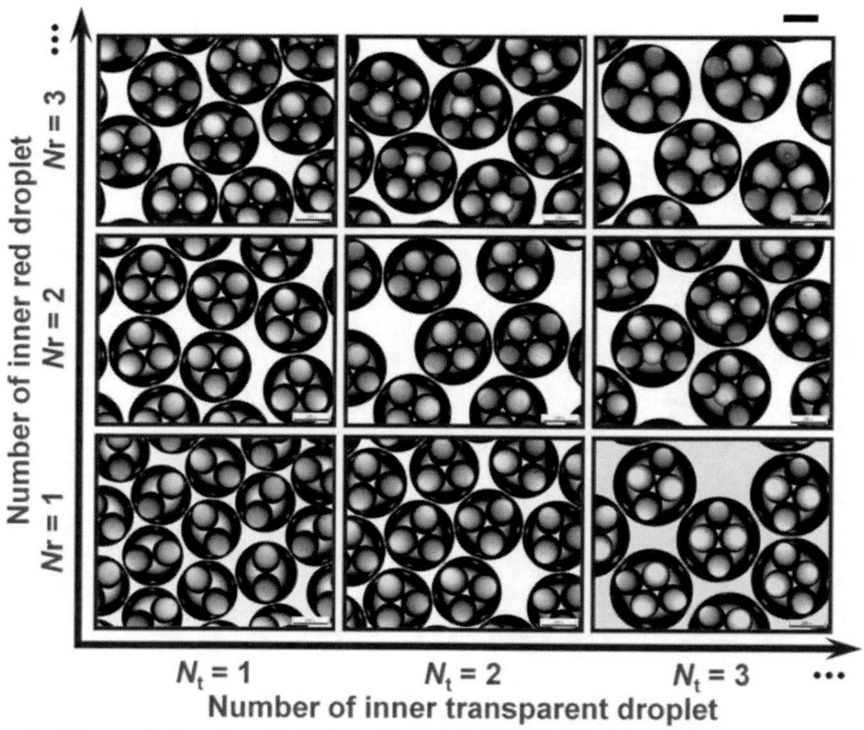

(Reproduced with permission from Ref.[30]). Copyright (2011), The Royal Society of Chemistry.

Figure 10. Optical micrographs of quadruple-component double emulsions exhibiting the precise control over the number and ratio of inner red and transparent droplets. Scale bars are 200 μm.

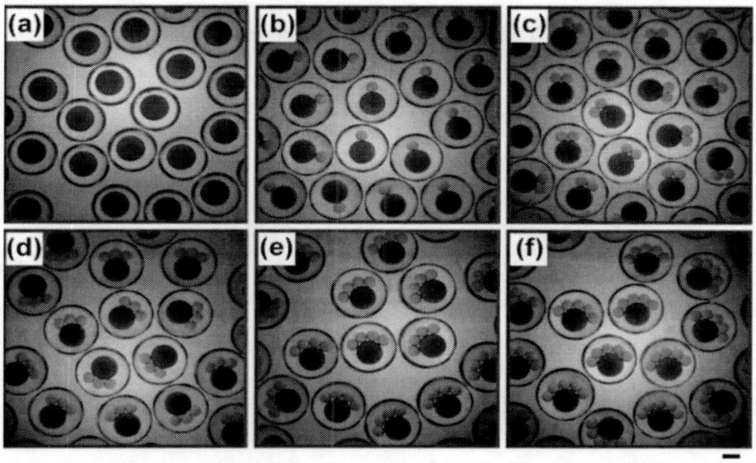

(Reproduced with permission from Ref.[30]). Copyright (2011), The Royal Society of Chemistry.

Figure 11. Optical micrographs showing controlled production of quadruple-component double emulsions with the number of smaller red droplets increasing from 0 to 5 and that of larger blue droplets being fixed at 1. Scale bar is 200 μm.

3.1.3. Independent Control of Connector and Liquid Extractor Over the Size of the Outer Droplets in Quadruple-Component Double Emulsions

Precise and independent control over the size of outer droplets can also be achieved using the functionality of the connectors and liquid extractors. Besides merging the main channel with branch channels, the connector is also designed to adjust the distance between droplets in the main channel by injecting additional fluid. When we increase the flow rate ($Q_{2\text{-C}}$) of injecting fluid ($F_{2\text{-C}}$) into the connector while keep the other flow rates constant, the distance between droplets increases as does the size of the outer droplets (d_3), due to the encapsulation of more fluid in the middle aqueous layer. This size control of outer droplets is shown in Figure 12a.

When the injecting flow rate ratio $Q_{2\text{-C}}/(Q_{1\text{-}1} + Q_{2\text{-}1} + Q_{1\text{-}2} + Q_{2\text{-}2} - Q_{2\text{-E}})$ increases from ~0.250 to ~1.333, the volume ratio V_i/V_0 of outer droplets increases from 1 to ~1.683. Importantly, in spite of the increase in d_3, the size of inner red droplets (d_1) and transparent droplets (d_2) remains unchanged, because the inner droplets are individually formed in branch channels.

By contrast, when we increase the flow rate through the liquid extractor ($Q_{2\text{-E}}$) while keeping the other flow rates constant, the distance between droplets decreases and the outer droplets become smaller due to the removal of continuous fluid. Increasing the extracting flow rate ratio $Q_{2\text{-E}}/(Q_{1\text{-}1} + Q_{2\text{-}1} + Q_{1\text{-}2} + Q_{2\text{-}2} + Q_{2\text{-C}})$ from ~0.034 to ~0.241 results in a decrease of volume ratio V_i/V_0 of outer droplets from 1 to ~0.736, with d_1 and d_2 unchanged (Figure 12b).

Because the data in Figure 12 are measured by using several batches of emulsion samples, the error bars mean that there is a little difference between the droplet diameters from batch to batch. However, the dripping mechanism for droplet formation offers highly control of the droplets, so the emulsions are still highly monodisperse in each batch, *i.e.*, the droplet formation is still well-controlled. These results illustrate the precise and independent size control over the outer droplets of a quadruple-component double emulsion.

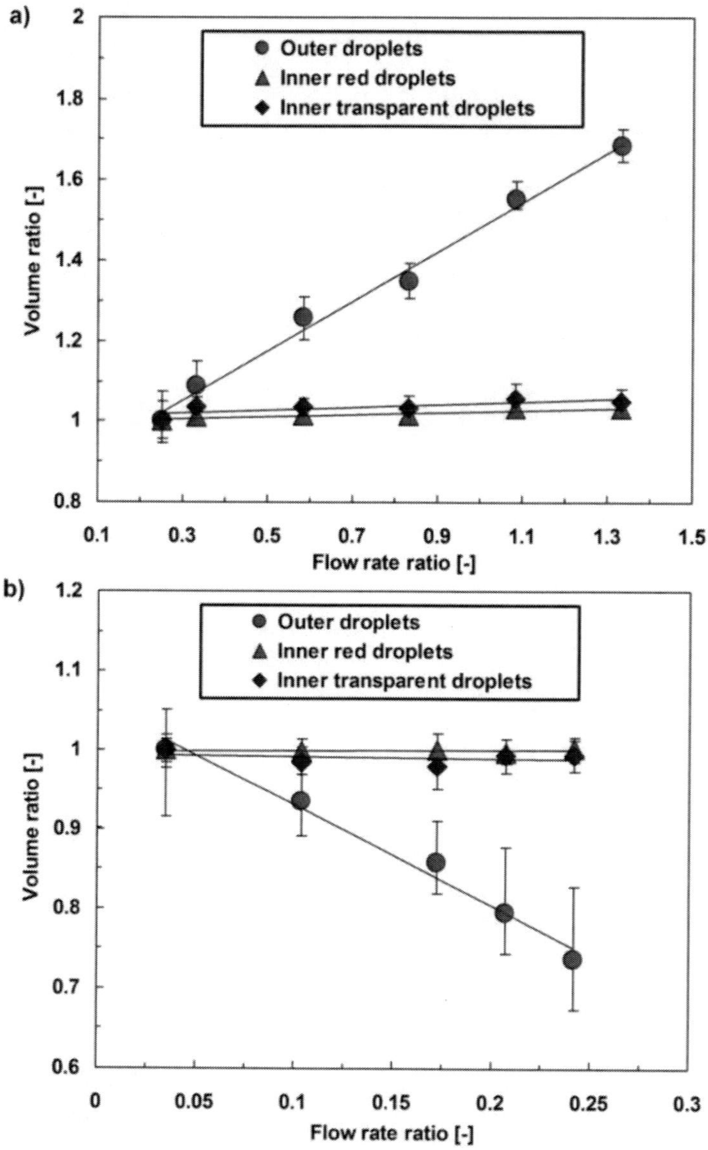

Figure 12. Effects of the flow rates of connector (a) and liquid extractor (b) on the sizes of inner and outer droplets in quadruple-component double emulsions. (a) Droplet volume ratio V_i/V_0 as a function of flow rate ratio $Q_{2-C}/(Q_{1-1} + Q_{2-1} + Q_{1-2} + Q_{2-2} - Q_{2-E})$. The initial values of d_1, d_2 and d_3 are 163.2 μm, 163.7 μm and 353.1 μm, respectively. (b) Droplet volume ratio V_i/V_0 as a function of flow rate ratio $Q_{2-E}/(Q_{1-1} + Q_{2-1} + Q_{1-2} + Q_{2-2} + Q_{2-C})$. V_0 is the initial droplet volume, and V_i is the droplet volume after varying Q_{2-C} (a) and Q_{2-E} (b). The initial values of d_1, d_2 and d_3 are 169.6 μm, 168.5 μm and 386 μm, respectively.

3.1.4. Prediction of the Inner Structure of the Quadruple-Component Double Emulsions

The excellent control over the number, ratio and size of co-encapsulated droplets allow us to quantitatively predict the inner structure of quadruple-component double emulsions generated from these microfluidic devices. For quadruple-component double emulsions, the total number (N) of both inner red and transparent droplets is controlled by matching the formation rates of inner red ($f_{1\text{-}1}$) and transparent droplets ($f_{1\text{-}2}$) with that of the outer droplets (f_3). According to mass conservation, the formation rate of each droplet can be described as follows:

$$f_{1-1} = \frac{Q_{1-1}}{\pi d_1^{\,3}/6} \tag{3}$$

$$f_{1-2} = \frac{Q_{1-2}}{\pi d_2^{\,3}/6} \tag{4}$$

$$f_3 = \frac{Q_3}{\pi d_3^{\,3}/6} \tag{5}$$

where $Q_{i\text{-}j}$ is the flow rate of fluid $F_{i\text{-}j}$; Q_3 is the sum of $Q_{1\text{-}1}$, $Q_{1\text{-}2}$, $Q_{2\text{-}1}$, $Q_{2\text{-}2}$, $Q_{2\text{-}C}$ and $Q_{2\text{-}E}$; d_i is the droplet diameter (as shown in Figure 8). For fixed device dimensions (D_i) and solution conditions, the droplet diameter (d_i) in coaxial flow depends on the velocity of the surrounding flow in the dripping regime [29], which can be described as follows:

$$\frac{d_1}{D_1} = a_1 \frac{Q_{1-1}}{Q_{2-1}} + b_1 \tag{6}$$

$$\frac{d_2}{D_2} = a_2 \frac{Q_{1-2}}{Q_{2-2}} + b_2 \tag{7}$$

$$\frac{d_3}{D_3} = a_3 \frac{Q_3}{Q_{3-\mathrm{M}}} + b_3 \tag{8}$$

where coefficients a_i and b_i are the slopes and intercepts that are obtained experimentally from these linear relations (Equations (6)-(8)) [29].

Based on Equations (2) to (8), we can quantitatively predict the value of N using the following equation:

$$N = \frac{f_{1-1} + f_{1-2}}{f_3} = \frac{\dfrac{Q_{1-1}}{D_1^3\left(a_1\left(Q_{1-1}/Q_{2-1}\right)+b_1\right)^3} + \dfrac{Q_{1-2}}{D_2^3\left(a_2\left(Q_{1-2}/Q_{2-2}\right)+b_2\right)^3}}{\dfrac{Q_3}{D_3^3\left(a_3\left(Q_3/Q_{3-\mathrm{M}}\right)+b_3\right)^3}} \tag{9}$$

The ratio of inner red droplets to transparent droplets (R_{12}) depends on the number of each kind of droplet encapsulated in the double emulsion; this can be separately controlled by adjusting $f_{1\text{-}1}$ and $f_{1\text{-}2}$. Therefore, we can quantitatively determine the value of R_{12} using the following equation:

$$R_{12} = \frac{f_{1-1}}{f_{1-2}} = \frac{\dfrac{Q_{1-1}}{\pi d_1^{\ 3}/6}}{\dfrac{Q_{1-2}}{\pi d_2^{\ 3}/6}} = \frac{Q_{1-1}}{Q_{1-2}} \frac{D_2^3\left(a_2\dfrac{Q_{1-2}}{Q_{2-2}}+b_2\right)^3}{D_1^3\left(a_1\dfrac{Q_{1-1}}{Q_{2-1}}+b_1\right)^3} \tag{10}$$

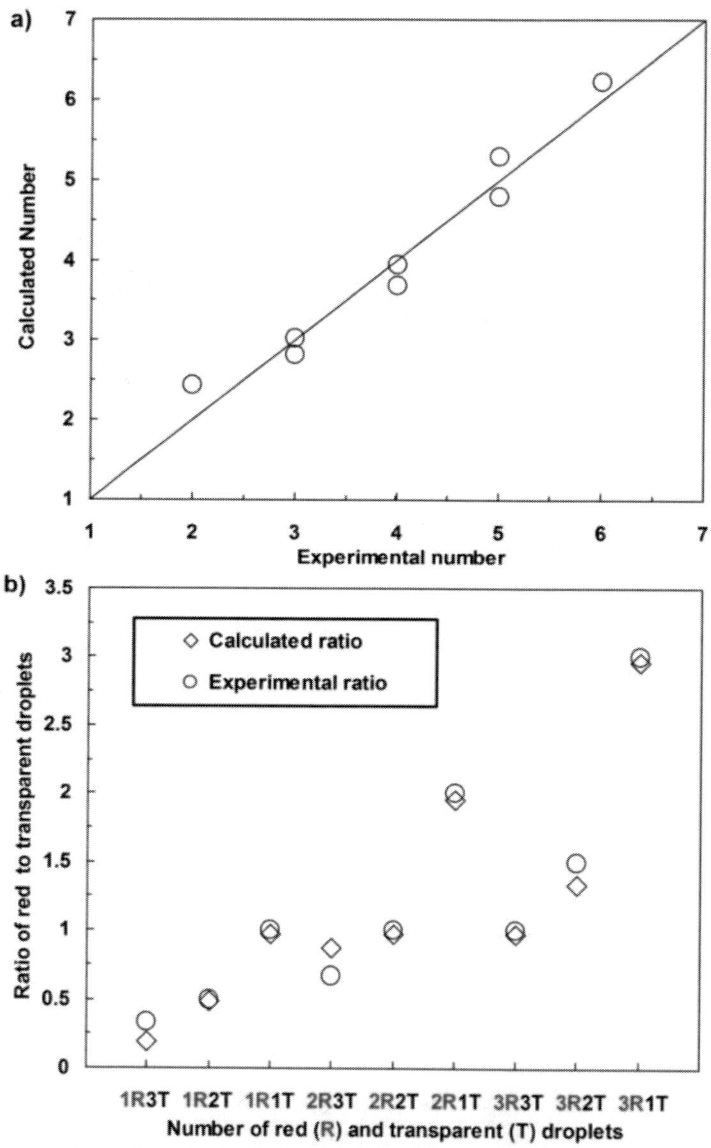

Figure 13. Prediction for the number (N) (a) and ratio (R_{12}) (b) of inner red and transparent droplets in quadruple-component double emulsions. (a) Prediction for the value of N, where the calculated value of N is obtained from Equation (9). (b) Prediction for the value of R_{12}, where the calculated value of R_{12} is obtained from Equation (10). For the device and quadruple-component double emulsions respectively shown in Figure 8 and Figure 10, the values of coefficients in Equations (9) and (10) are a_1=0.3957, b_1=0.9742, a_2=0.2482, b_2=1.0137, a_3=0.8451, and b_3=0.4894.

The calculated values of N and R_{12} from Equations (9) and (10) are compared with the experimental values in Figure 13. The good agreement between the calculated and experimental values of both N and R_{12} indicates that we can successfully predict the inner structure of quadruple-component double emulsions. By exploiting mass conservation of each fluid, further equations can be developed to predict the inner structure of higher-order multicomponent multiple emulsions.

3.2. Scale-Up of Device for Higher-Order Multicomponent Multiple Emulsions

The variety in combinations of the building blocks used in these devices enables simple scale-up for controlled generation of higher-order multicomponent multiple emulsions with more complex structures. This is illustrated by controllably fabricating quintuple-component double emulsions, quintuple-component triple emulsions, and even sextuple-component triple emulsions.

3.2.1. Controllable Monodisperse Quintuple-Component Double Emulsions

For generation of quintuple-component double emulsions, we scale up the microfluidic device in Figure 8 by adding an additional branch channel upstream of the main channel to introduce a third kind of oil droplets into the quadruple-component double emulsions, as shown in Figure 14a. These additional droplets are dyed with Sudan black, so the quintuple-component O/W/O double emulsions contain oil droplets of three colors. Micrographs of the quintuple-component double emulsions, where the number of each kind of droplets increases from 1 to 5, are shown in Figure 14b.

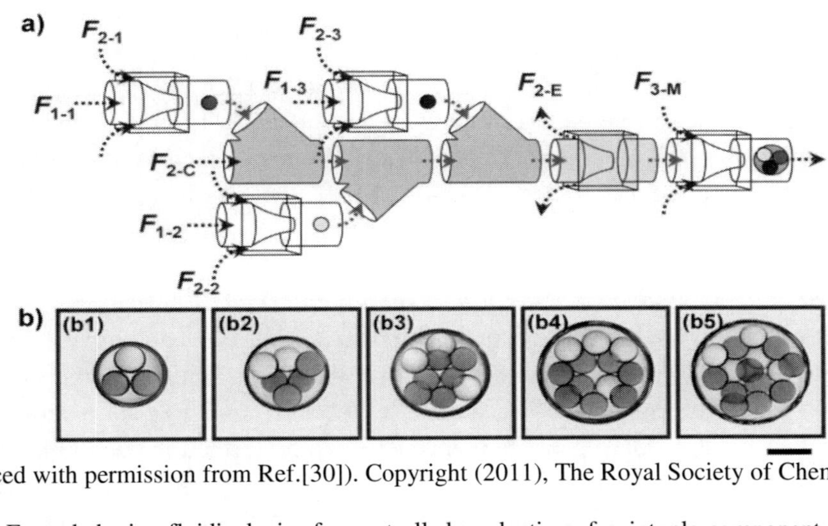

Figure 14. Extended microfluidic device for controlled production of quintuple-component double emulsions. (a) Schematic diagram of microfluidic device for generating quintuple-component double emulsions. (b) Optical micrographs of quintuple-component double emulsions that contain three different kinds of inner droplets, with a controlled increase in the number of each droplet increases from 1 to 5. Scale bars are 200 μm.

3.2.2. Controllable Monodisperse Quintuple-Component Triple Emulsions

For fabrication of quintuple-component triple emulsions, an additional drop maker is added downstream in the main channel of the microfluidic device in Figure 8 to further emulsify the quadruple-component double emulsions, as shown in Figure 15a.

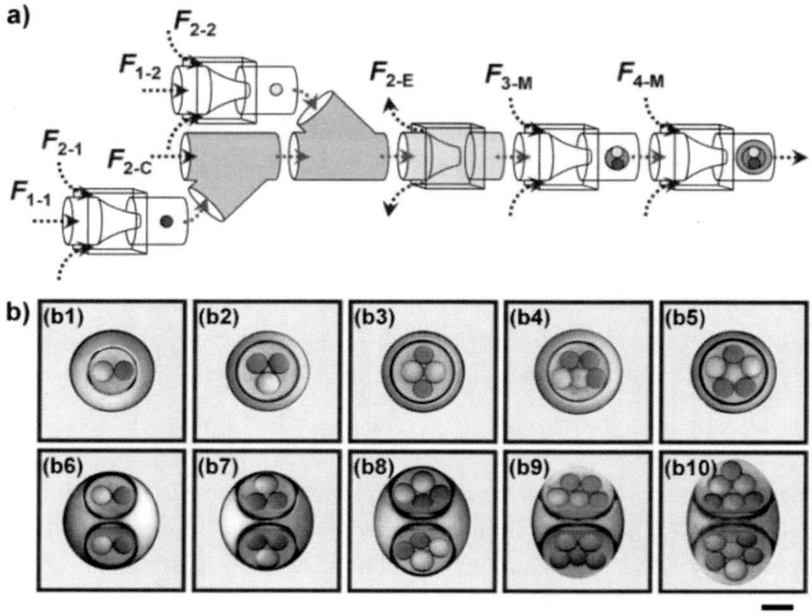

Figure 15. Extended microfluidic device for controlled production of quintuple-component multiple emulsions. (a) Schematic diagram of microfluidic device for generating quintuple-component triple emulsions. (b) Optical micrographs of quintuple-component triple emulsions that contain two different kinds of droplets in the innermost level, with precise control over the number and ratio of different innermost droplets and over the number of the middle droplets. Scale bars are 200 μm.

The quintuple-component oil-in-water-in-oil-in-water (O/W/O/W) triple emulsions are generated with red and transparent oil droplets in the innermost level surrounded by a blue aqueous layer, which is further envelopped in an oil droplet. The number and ratio of innermost red and transparent oil droplets can be precisely controlled as can the number of blue aqueous drops (Figure 15b).

3.2.3. Controllable Monodisperse Sextuple-Component Triple Emulsions

For generation of more complex sextuple-component triple emulsions, we place two drop makers in each branch channel for the microfluidic device in Figure 8 to generate double emulsions, as shown in Figure 16a. Because $F_{1\text{-}1}$, $F_{2\text{-}1}$, $F_{1\text{-}2}$ and $F_{2\text{-}2}$ are each of different components, two kinds of O/W/O double emulsions can be formed in each branch channel, each differing in compositions of both inner and middle phases. After an additional encapsulation of the double emulsions in the main channel, highly controllable and monodisperse sextuple-component triple emulsions are generated, as shown in Figure 16b. These uniform O/W/O/W triple emulsions contain two different kinds of double emulsions inside, one of which contains red oil droplets in each transparent aqueous droplet while the other of which contains transparent oil droplets in each blue aqueous droplet (Figures 16(b1) and 16(b2)). Interestingly, when we stop injecting one of the innermost fluid (F_{1-2}), a novel type of triple emulsion, which has both double and single emulsions co-encapsulated in the same level, is generated (Figure 16(b3)).

Even with such complex structures, precise control over the inner structure of sextuple-component triple emulsions is still achievable (Figure 17). The number and ratio of the transparent and blue aqueous droplets in the second inner level can be precisely controlled by matching the formation rates of the relavant fluids (Figure 17, from row I to row IV).

By selectively stopping the injection of either of the innermost fluids, we can produce sextuple-component triple emulsions with no red oil droplet in the encapsulated transparent aqueous droplets (Figure 17, column 1), or with no transparent oil droplet in the encapsulated blue aqueous droplets (Figure 17, columns 2 and 3). We can also achieve precise control over the number of different innermost droplets as well as the ratio of innermost transparent to red droplets (Figure 17, columns 4 to 6). All higher-order multicomponent multiple emulsions highlight the remarkable scalability of our microfluidic device; this is crucial because it enables us to arrange the building blocks to engineer multicomponent multiple emulsions with specified complex structures.

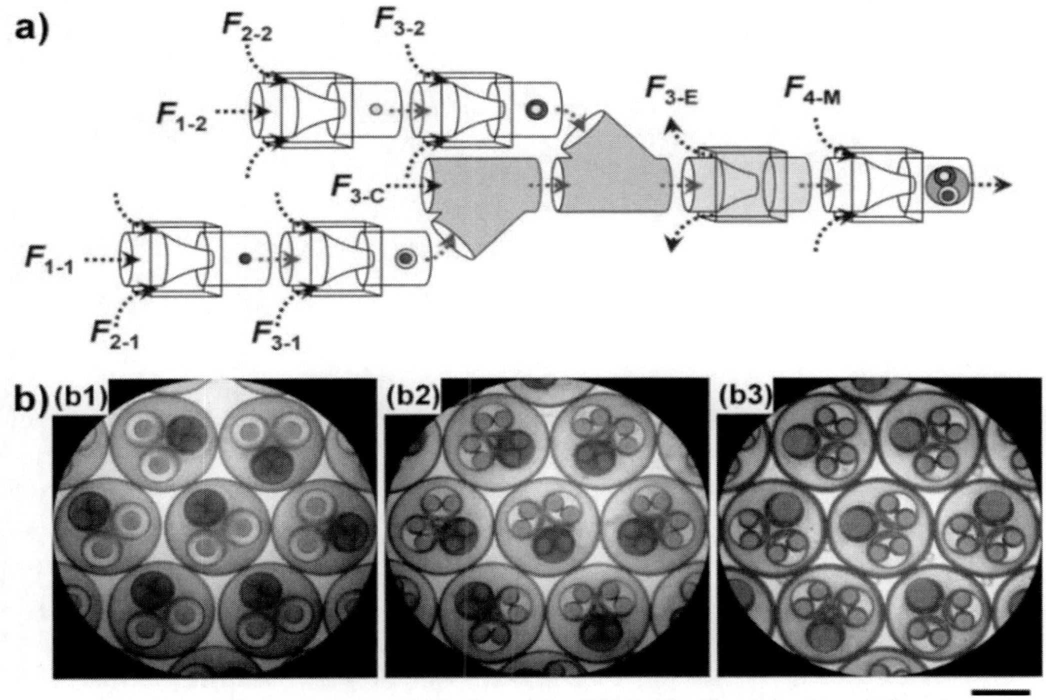

Figure 16. Extended microfluidic device for controlled production of sextuple-component triple emulsions. (a) Schematic diagram of microfluidic device for generating sextuple-component triple emulsions. (b) Optical micrographs of monodisperse sextuple-component triple emulsions. (b1, b2) O/W/O/W triple emulsions containing two different double emulsions and (b3) novel triple emulsions containing one W/O single emulsion and two O/W/O double emulsions. Scale bars are 400 μm.

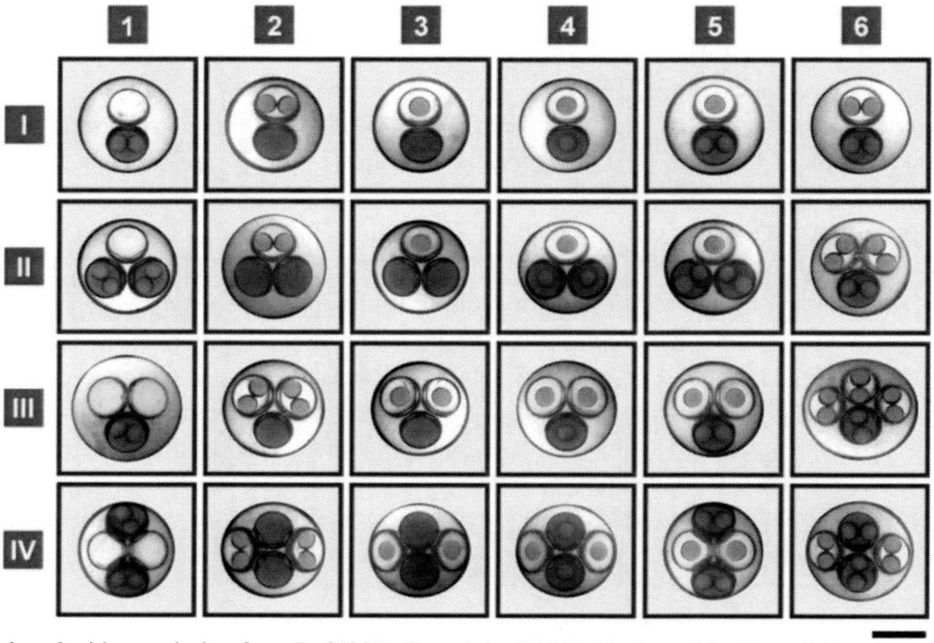

(Reproduced with permission from Ref.[30]). Copyright (2011), The Royal Society of Chemistry.

Figure 17. Optical micrographs of sextuple-component triple emulsions showing the precise control over the number and ratio of different droplets co-encapsulated in each level of O/W/O/W triple emulsions. Scale bars are 400 μm.

4. Microfluidic Fabrication of Novel Materials from Multiple Emulsions

The ability to precisely control the formation of multiple emulsions offers new opportunities to engineer novel materials. To illustrate this potential, we use the triple emulsion structure to fabricate a microcapsule made with a thermo-sensitive, poly(*N*-isopropylacrylamide) hydrogel that can be used for controlled release of actives. Since each fluid layer is sandwiched between two immiscible fluids, we are able to perform a polymerization reaction in a specific layer.

For the thermo-sensitive microcapsules, here we form a W/O/W/O triple emulsion, and add monomer, crosslinker, and an initiator in the outer aqueous layer. We used the same outermost and innermost fluids as for the triple emulsions. The outer middle fluid (II) was an aqueous solution containing 10 wt% glycerol, 2 % (w/v) PVA, 11.3 % (w/v) of the monomer *N*-isopropylacrylamide (NIPAM), 0.8 % (w/v) of the co-monomer sodium acrylate, 0.77 % (w/v) of the crosslinker *N,N'*-methylenebisacryamide (BIS), and 0.6 % (w/v) of the initiator ammonium persulfate (APS). The inner middle fluid (I) was 10 cSt PDMS oil containing 5 wt% Dow Corning 749 fluid and 8 % (v/v) of the accelerator *N,N,N',N'*-tetramethylethylenediamine (TEMED). The accelerator TEMED is both oil- and water-soluble.

When TEMED is dissolved in the middle fluid (I), it diffuses into the outer middle fluid (II) where it meets the initiator APS; this starts a redox reaction that polymerizes the monomers. The result is a microcapsule consisting of a shell of thermo-sensitive hydrogel which encapsulates an oil drop containing several water droplets, all in a continuous oil phase, as shown in Figure 18a. The polymerized microcapsule together with the surrounding fluids was contained in a transparent holder on a slide glass, which was placed on a microscope-mounted heating and cooling stage to measure the thermo-sensitive behavior. A digital camera recorded the behavior and the release process. Upon heating from 25° to 50° C, the thermo-sensitive hydrogel rapidly shrinks by expelling water; however, because of the incompressibility of the inner oil, the hydrogel shell breaks, providing spontaneous, pulsed release of the innermost water droplets into the continuous oil phase, as shown in Figure 18b-e. This structure has a Trojan-horse-like behavior, protecting the innermost water droplets in the hydrogel shell until their temperature-induced release. This demonstrates the utility of this technique to generate highly controlled capsules with multiple internal volumes that remain separate from each other; it also highlights the potential of this device to create highly engineered structures for controlled release of actives. Further refinements could adjust the thickness of the layers and the number of droplets, enabling fine control over diffusion of actives, contained within the innermost droplets. This would facilitate highly controlled release of the actives.

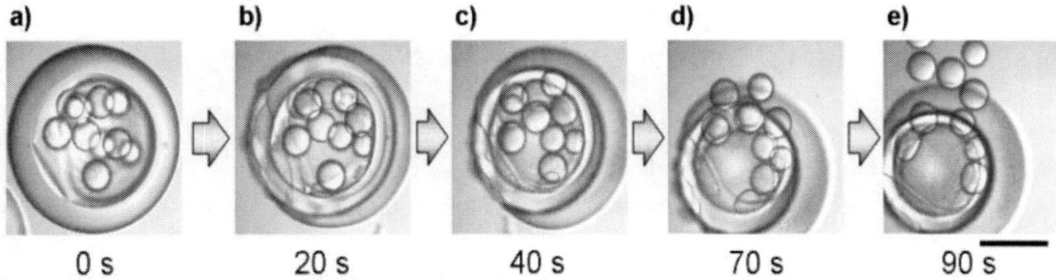

Figure 18. Temperature sensitive hydrogel microcapsule for pulsed release. (a) Optical micrograph of a microcapsule with a shell comprised of a thermo-sensitive hydrogel containing aqueous droplets dispersed in oil. Upon increase of the temperature, the hydrogel shell shrinks by expelling water. This capsule was generated from a triple emulsion, where the continuous liquid is oil, the hydrogel shell is aqueous, the inner middle fluid is also oil, and the innermost droplets are aqueous. (b-e) Optical micrograph time-series showing the forced expulsion of the oil and water droplets contained within the microcapsule when the temperature is rapidly increased from 25 to 50 °C. The time-series begins once the temperature reaches 50 °C. The appearance of the extra layer surrounding the microcapsule in (b-e) is water that is squeezed out from the hydrogel shell as it shrinks. The coalescence of the expelled inner oil with the outer oil can not be resolved since both liquids have the same index of refraction. The scale bar is 200 μm.

SUMMARY AND PERSPECTIVE

The excellent controllability and remarkable scalability afforded by these capillary microfluidic techniques makes it a flexible and promising route for engineering highly controlled multiple emulsions with diverse structures. The approach presented here circumvents the difficulties in controllably forming multi-phase structures. It can be used to optimize multiple emulsion systems for a broad range of applications in pharmaceutics, foods, cosmetics and separations. Moreover, its generality will enable fabrication of novel materials containing complex internal structures. Future work must focus on refining methods proposed here to allow the full potential of the technique to be realized. In addition, practical applications of these devices would require operating a large number of devices in parallel, which might best be accomplished through other fabrication techniques. Nevertheless, the results presented here provide an essential guide to assess the utility of the structures that can be produced.

REFERENCES

[1] Davis, S. S. and Walker, I. M. (1987). Multiple emulsions as targetable delivery systems. *Methods in enzymology, 149,* 51-64.

[2] Nakano, M. (2000). Places of emulsions in drug delivery. *Advanced Drug Delivery Reviews, 45(1)*, 1-4.

[3] Vasiljevic, D., Parojcic, J., Primorac, M. and Vuleta, G. (2006). An investigation into the characteristics and drug release properties of multiple W/O/W emulsion systems containing low concentration of lipophilic polymeric emulsifier. *International Journal of Pharmaceutics, 309(1-2)*, 171-177.

[4] Weiss, J., Scherze, I. and Muschiolik, G. (2005). Polysaccharide gel with multiple emulsion. *Food Hydrocolloids, 19(3)*, 605-615.

[5] Lobato-Calleros, C., Rodriguez, E., Sandoval-Castilla, O., Vernon-Carter, E. J. and Alvarez-Ramirez, J. (2006). Reduced-fat white fresh cheese-like products obtained from W-1/O/W-2 multiple emulsions: Viscoelastic and high-resolution image analyses. *Food Research International, 39(6)*, 678-685.

[6] Yoshida, K., Sekine, T., Matsuzaki, F., Yanaki, T. and Yamaguchi, M. (1999). Stability of vitamin A in oil-in-water-in-oil-type multiple emulsions. *Journal of the American Oil Chemists Society, 76(2)*, 195-200.

[7] Lee, M. H., Oh, S. G., Moon, S. K. and Bae, S. Y. (2001). Preparation of silica particles encapsulating retinol using O/W/O multiple emulsions. *Journal of Colloid and Interface Science, 240(1)*, 83-89.

[8] Chakravarti, A. K., Chowdhury, S. B. and Mukherjee, D. C. (2000). Liquid membrane multiple emulsion process of separation of copper(II) from waste waters. *Colloids and Surfaces a-Physicochemical and Engineering Aspects, 166(1-3)*, 7-25.

[9] Chakraborty, M., Ivanova-Mitseva, P. and Bart, H.-J. (2006). Selective separation of toluene from n-heptane via emulsion liquid membranes containing substituted cyclodextrins as carrier. *Separation Science and Technology, 41(16)*, 3539-3552.

[10] Utada, A. S., Lorenceau, E., Link, D. R., Kaplan, P. D., Stone, H. A. and Weitz, D. A. (2005). Monodisperse double emulsions generated from a microcapillary device. *Science, 308(5721)*, 537-541.

[11] Lorenceau, E., Utada, A. S., Link, D. R., Cristobal, G., Joanicot, M. and Weitz, D. A. (2005). Generation of polymerosomes from double-emulsions. *Langmuir, 21(20)*, 9183-9186.

[12] Nie, Z. H., Xu, S. Q., Seo, M., Lewis, P. C. and Kumacheva, E. (2005). Polymer particles with various shapes and morphologies produced in continuous microfluidic reactors. *Journal of the American Chemical Society, 127(22)*, 8058-8063.

[13] Bocanegra, R., Sampedro, J. L., Ganan-Calvo, A. M. and Marquez, M. (2005). Monodisperse structured multi-vesicle microencapsulation using flow-focusing and controlled disturbance. *Journal of Microencapsulation, 22(7)*, 745-759.

[14] Rizkalla, N., Range, C., Lacasse, F. X. and Hildgen, P. (2006). Effect of various formulation parameters on the properties of polymeric nanoparticles prepared by multiple emulsion method. *Journal of Microencapsulation, 23(1)*, 39-57.

[15] Zoldesi, C. I. and Imhof, A. (2005). Synthesis of monodisperse colloidal spheres, capsules, and microballoons by emulsion templating. *Advanced Materials, 17(7)*, 924-928.

[16] Koo, H. Y., Chang, S. T., Choi, W. S., Park, J. H., Kim, D. Y. and Velev, O. D. (2006). Emulsion-based synthesis of reversibly swellable, magnetic nanoparticle-embedded polymer microcapsules. *Chemistry of Materials, 18(14)*, 3308-3313.

[17] Nakashima, T., Shimizu, M. and Kukizaki, M. (2000). Particle control of emulsion by membrane emulsification and its applications. *Advanced Drug Delivery Reviews, 45(1)*, 47-56.

[18] Vladisavljevic, G. T., Shimizu, M. and Nakashima, T. (2006). Production of multiple emulsions for drug delivery systems by repeated SPG membrane homogenization: Influence of mean pore size, interfacial tension and continuous phase viscosity. *Journal of Membrane Science, 284(1-2)*, 373-383.

[19] Sugiura, S., Nakajima, M., Yamamoto, K., Iwamoto, S., Oda, T., Satake, M. and Seki, M. (2004). Preparation characteristics of water-in-oil-in-water multiple emulsions using microchannel emulsification. *Journal of Colloid and Interface Science, 270(1)*, 221-228.

[20] Okushima, S., Nisisako, T., Torii, T. and Higuchi, T. (2004). Controlled production of monodisperse double emulsions by two-step droplet breakup in microfluidic devices. *Langmuir, 20(23)*, 9905-9908.

[21] Loscertales, I. G., Barrero, A., Guerrero, I., Cortijo, R., Marquez, M. and Ganan-Calvo, A. M. (2002). Micro/nano encapsutation via electrified coaxial liquid jets. *Science, 295(5560)*, 1695-1698.

[22] Chen, H., Zhao, Y., Song, Y. and Jiang, L. (2008). One-step multicomponent encapsulation by compound-fluidic electrospray. *Journal of the American Chemical Society, 130(25)*, 7800-7801.

[23] Pays, K., Giermanska-Kahn, J., Pouligny, B., Bibette, J. and Leal-Calderon, F. (2002). Double emulsions: how does release occur? *Journal of Controlled Release, 79(1-3)*, 193-205.

[24] Mason, T. G. and Bibette, J. (1997). Shear rupturing of droplets in complex fluids. *Langmuir, 13(17)*, 4600-4613.

[25] Schmitt, V., Leal-Calderon, F. and Bibette, J. (2003). Preparation of monodisperse particles and emulsions by controlled shear. *Topics in Current Chemistry, 227*,195-215.

[26] Pays, K., Giermanska-Kahn, J., Pouligny, B., Bibette, J. and Leal-Calderon, F. (2001). Double emulsions: A tool for probing thin-film metastability. *Physical Review Letters, 87(17)*, 178304.

[27] Goubault, C., Pays, K., Olea, D., Gorria, P., Bibette, J., Schmitt, V. and Leal-Calderon, F. (2001). Shear rupturing of complex fluids: Application to the preparation of quasi-monodisperse water-in-oil-in-water double emulsions. *Langmuir, 17(17)*, 5184-5188.

[28] Nisisako, T., Okushima, S. and Torii, T. (2005). Controlled formulation of monodisperse double emulsions in a multiple-phase microfluidic system. *Soft Matter, 1(1)*, 23-27.

[29] Chu, L. Y., Utada, A. S., Shah, R. K., Kim, J. W. and Weitz, D. A. (2007). Controllable monodisperse multiple emulsions. *Angewandte Chemie-International Edition, 46(47)*, 8970-8974.

[30] Wang, W., Xie, R., Ju, X. J., Luo, T., Liu, L., Weitz, D. A. and Chu, L. Y. (2011). Controllable microfluidic production of multicomponent multiple emulsions. *Lab on a Chip, 11(9)*, 1587-1592.

[31] Manoharan, V. N., Elsesser, M. T. and Pine, D. J. (2003). Dense packing and symmetry in small clusters of microspheres. *Science, 301(5632)*, 483-487.

[32] Umbanhowar, P. B., Prasad, V. and Weitz, D. A. (2000). Monodisperse emulsion generation via drop break off in a coflowing stream. *Langmuir, 16(2)*, 347-351.

[33] Sun, B. J., Shum, H. C., Holtze, C. and Weitz, D. A. (2010). Microfluidic Melt Emulsification for Encapsulation and Release of Actives. *ACS Applied Materials and Interfaces, 2(12)*, 3411-3416.

[34] Abate, A. R. and Weitz, D. A. (2009). High-Order Multiple Emulsions Formed in Poly(dimethylsiloxane) Microfluidics. *Small, 5(18)*, 2030-2032.

[35] Pannacci, N., Bruus, H., Bartolo, D., Etchart, I., Lockhart, T., Hennequin, Y., Willaime, H. and Tabeling, P. (2008). Equilibrium and Nonequilibrium States in Microfluidic Double Emulsions. *Physical Review Letters, 101(16)*, 164502.

[36] Bauer, W.-A. C., Fischlechner, M., Abell, C. and Huck, W. T. S. (2010). Hydrophilic PDMS microchannels for high-throughput formation of oil-in-water microdroplets and water-in-oil-in-water double emulsions. *Lab on a Chip, 10(14)*, 1814-1819.

[37] Seo, M., Paquet, C., Nie, Z., Xu, S. and Kumacheva, E. (2007). Microfluidic consecutive flow-focusing droplet generators. *Soft Matter, 3(8)*, 986-992.

[38] Atencia, J. and Beebe, D. J. (2005). Controlled microfluidic interfaces. *Nature, 437(7059)*, 648-655.

[39] Whitesides, G. M. (2006). The origins and the future of microfluidics. *Nature, 442(7101)*, 368-373.

In: Microfluidics
Editors: Stefano Panzarella and Walter Maroni

ISBN: 978-1-62417-345-5

Chapter 4

AN ALTERNATIVE METHOD TO PREDICT WETTABILITY FOR MICROFLUIDICS

Li Yang [*] ***and Alex G. Shard***
Analytical Science Division, National Physical Laboratory,
Teddington, Middlesex, UK

ABSTRACT

Wettability properties inside the microchannels of microfluidic devices have received increasing attention over the past few years due to their important roles in controlling liquid flow. Conventionally, the wetting properties of a surface are determined by contact angle measurement where a liquid drop is placed on the surface and the angle between the solid surface and liquid-vapour interface is measured. The size of the liquid drop on the surface is often larger than the size of components or features in microfluidic systems. This renders conventional methods for measuring contact angle incapable of delivering the information required. This chapter presents a brief review of our recent research in the development of an indirect approach for predicting contact angle by the use of surface analytical methods such as time-of flight secondary ion mass spectrometry (ToF-SIMS) and x-ray photoelectron spectroscopy (XPS). Two examples have been shown here. One is to predict micron-scale surface wettability maps of substrates by exploitation of surface chemical imaging using SIMS. Using the surface sensitive and chemically specific spectroscopy available with SIMS, it was possible to obtain information on species that are related to wetting behavior and their distribution over the surface. Another example has demonstrated that XPS, similar to SIMS, has the potential to give precise surface chemical information and provide insights into the quantitative assessment of the surface functional groups, allowing the prediction of surface wettability. The alternative method described here is of direct benefit to industries where wettability is crucial but contact angle measurements cannot be directly acquired on more complicated or flexible surfaces.

[*] Corresponding author: li.yang@npl.co.uk;

1. Introduction

Microfluidics deals with the behaviour, precise control and manipulation of fluids, which are geometrically constrained to a length scale less than a millimetre. The first microfluidic protocol can be traced back to the development of a silicon chip based gas chromatography at Stanford University [1] and the ink-jet printer at IBM [2] in the late 1970s. It was not until the early 1990s that the concept of "miniaturised total chemical analysis system" (or μTAS) was proposed by Manz et al [3]. Microfluidics has now become a multidisciplinary field involving engineering, physics, chemistry, microtechnology and biotechnology, with practical applications of DNA chips [4], cell sorting [5], lab-on-a-chip [6], inkjet-printheads [7], microfluidic reactors [8] and microchannels [9]. The enormous attention received by this subject has been driven by the ability to fabricate individual and integrated flow configurations in the scale order of microns and smaller. Other applications involve developments in biomedical diagnostics and biotechnology for which manipulations are required on the cellular length scale, and the potential use of microsystems to synthesise, formulate, process and analyse fluids.

There are many excellent reviews which provide an overview of techniques in microfluidics [10, 11, 12, 13, 14, 15]. The microfluidics literature describes many functional elements including valves, pumps, dispensers, mixers, filters, separators and heaters. Some of them contain new flow configurations, the lab-on-a-chip concept, reactions using only very small fluid volumes and microflow dynamics in a single channel. In this chapter, we do not cover every aspect of microfluidics, but emphasise the topic of providing an alternative method to predict wettability for microfluidic devices or other size-constrained microchannels. As dimensions shrink, there is a need to develop a method to assess the wetting properties (contact angle, surface tension) along the microchannel. We will focus on two examples from our own work, which have demonstrated an indirect approach for predicting wettability by the use of surface analytical methods: namely time-of flight secondary ion mass spectrometry (ToF-SIMS) [16] and x-ray photoelectron spectroscopy (XPS) [17]. By using appropriate reference surfaces for which the wettability can be correlated with surface chemical data from any of these analytical techniques, it is possible to provide calibration curves or plots to relate their wettability behavior to their surface chemical properties.

2. Basic Concepts for Wettability Prediction

The wettability of materials can be determined by contact angle measurement. However, conventional contact angle measurements are difficult on microchannels, microfluidic or plastic electronic devices, since the size of the liquid drop on the surface is often larger than their typical dimensions. Direct contact angle measurements on devices are therefore not straightforward. An indirect approach for predicting contact angle is of direct benefit to industries where wettability is crucial but contact angle measurements cannot be directly acquired.

Figure 1 shows the basic wetting configuration and the definition of the contact angle [18], θ, an angle between the solid surface and the tangent of the liquid-vapour interface of a sessile drop. The basic Young-Dupré equation defining the contact angle at equilibrium is [19]

$$\gamma_{SG} = \gamma_{SL} + \gamma_{LG} \cos\theta \tag{1}$$

where, as defined in Figure 1, γ_{SG} is the surface energy between the solid substrate and the gas (G) or air environment, γ_{LG} is the surface energy of the water droplet and γ_{SL} is the interfacial energy of the solid substrate under the liquid (L) or water. Changing the substrate surface alters both γ_{SG} and γ_{SL}, but not γ_{LG} unless soluble products are formed. Using simple pair-bond theory, both γ_{SG} and γ_{SL} will depend on the number of atoms in the surface of each species in any given chemical state. The use of static SIMS or XPS has the ability to provide compositional information over the outermost atom layer. Therefore, the quantitative data with detailed spectra signals makes it possible to relate to the surface compositions of the materials properties and interpolate properties for any system, e.g. microfluidics.

Baytekin and his co-workers [20] have demonstrated the use of surface analytical methods such as ToF-SIMS and XPS to study points of interest in microfluidic devices and provide information on the surface chemistry established there. By using flat reference polymer wafers made from the same material, the contact angle can be measured as usual, and appropriate calibration curves can be established. Analytical data obtained from real microchannels can then be evaluated in terms of wetting properties using the calibration curves. A later paper [21] by Baytekin and co-workers extends this work to hot embossed polymers poly (methyl methacrylate) (PMMA) and polycarbonate (PC) using both ToF-SIMS and XPS data correlated with contact angle measurements, over the limited data range used, to within 2° and 3°. The repeatability of ToF-SIMS measurements were investigated by principle component analysis (PCA) in some detail. PCA enables to separate and classify the ToF-SIMS spectra of plasma treated and untreated wafers and microchannel devices according to treatment conditions. Figure 2 shows the correlation between the cosine of the water contact angles before and after different plasma treatments and XPS O/C ratios of PMMA reference wafers. Highly active species within the oxygen plasma cause a surface chemistry change in PMMA samples. Low-molecular-weight polar materials will present on the surface due to plasma treatment. From this calibration curve, the wettability for a PMMA micro-channel in a micro-fluidic Faraday diode test structure was estimated as 55°. By carefully establishing calibration curves, information on surface chemistry as obtained by ToF-SIMS or XPS can be interpreted the wettability for the inner walls of micro-channels.

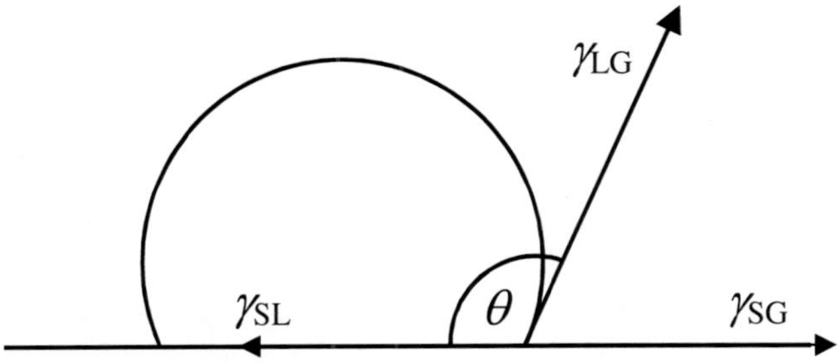

Figure 1. Schematic of the wetting configuration. Reprinted with permission from ref. [17]. Copyright © 2010 American Institute of Physics.

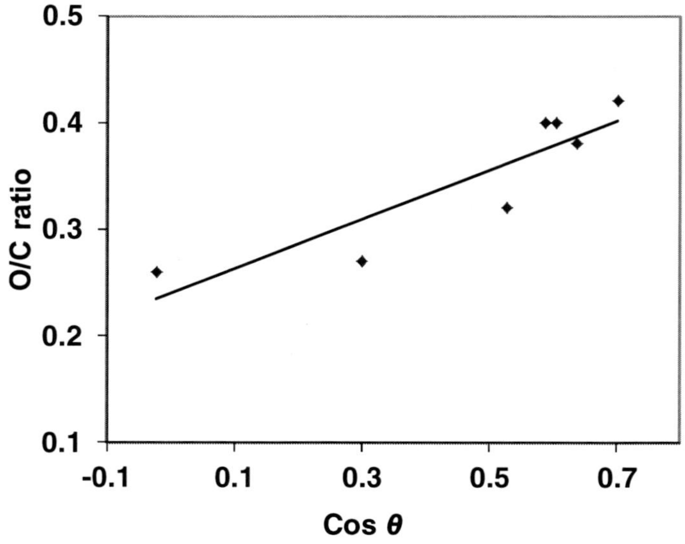

Figure 2. The correlation between the measured values of cosθ and XPS O/C surface concentration ratios measured for oxygen plasma treated PMMA wafers. Reproduced from the data in ref. [21]

Urquhart and co-workers [22] have also reported the use of ToF-SIMS analysis to determine the relationship between water contact angle (WCA) and the surface chemistry of a polymer library array using Partial Least Squares (PLS) regression [23]. A good correlation between ToF-SIMS data from the 576 acrylate-based polymers and contact angle within 6° (or 0.09 in cosθ) was obtained (as shown in Figure 3). Figure 3a displays a good linear fit of the model predicted WCA values against their measured values ranging between 47° and 93°. The resulting regression coefficients illustrated in Figure 3b give the insight of surface fragments with the relationship of WCA. Secondary ions with large positive regression coefficients, e.g. $C_nH_m^+$ are nonoxygenated hydrocarbon fragments, which are associated with higher WCA values. While ions with large negative regression coefficients are likely linked to lower WCA values due to linear oxygenated hydrocarbon fragments ($C_nH_mO^+$) and amine/amide fragments (CN^-, CNO^- and CH_4N^+). In further work from their group [24], PLS has been shown to be able to identify surface moieties, which are important in controlling surface energy in a 496 copolymer micro-patterned library. Their studies demonstrate that multivariate analysis can be successfully applied to ToF-SIMS data from a large library of samples and such an approach is a potential technique for building useful models of wettability from surface spectroscopies.

Gradowski used multivariate modelling to correlate ToF-SIMS spectra with the wetting behaviour of plasma-polymer films [25]. Figure 4 shows the application of PLS to various substrates (Si, PET and tissue material). The surfaces have been modified with CF_x groups and the contact angle for wetting by water has been measured. The correlation between ToF-SIMS intensities and contact angles resulting from a PLS model data used as a training set is shown with open markers. The solid markers demonstrate that this model can accurately predict results from further samples that were not originally included in the training set. The predicted and measured contact angles are excellently correlated with a scatter of 5%.

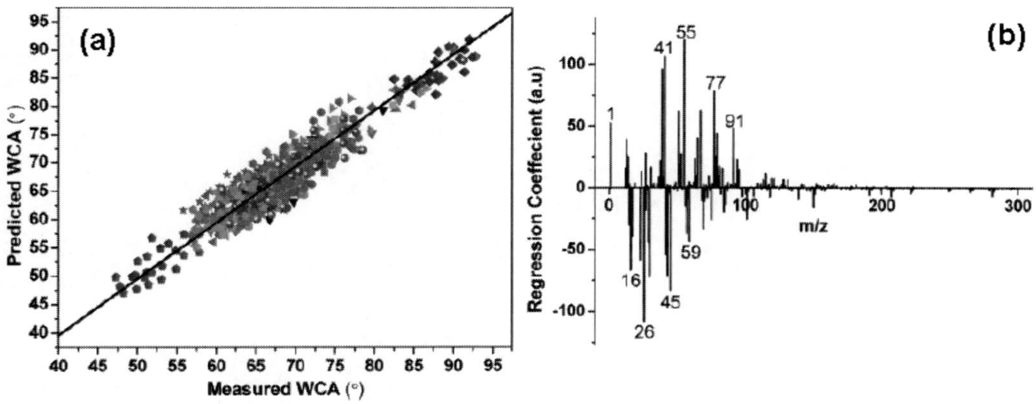

Figure 3. A PLS model showing the correlation between ToF-SIMS spectra and wettability of 576 polymers. (a) The predicted WCA against measured WCA. (b) The regression coefficients versus peak *m/z* for the data set, indicating high WCA correlating with polar, hydrophilic fragments and vice versa. Reprinted with permission from ref. [22]. Copyright © 2008 American Chemical Society.

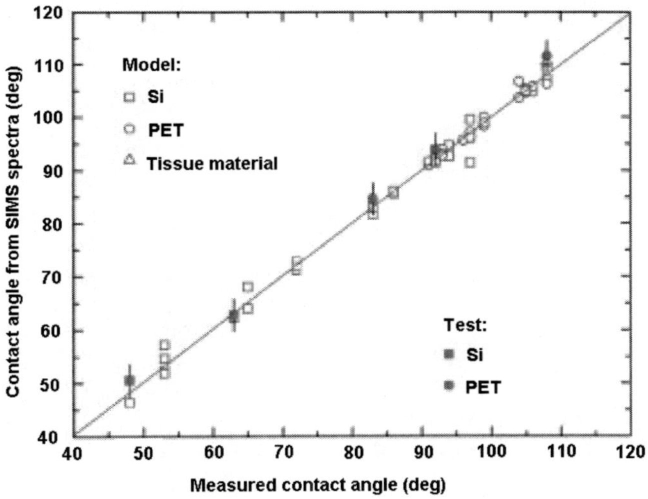

Figure 4. The correlation between measured contact angles and contact angles resulted from PLS calculations using model data. Reproduced from the data in ref. [25].

Paynter [26] showed that a linear negative correlation was found between the oxygen content of the polystyrene surface (determined from angle-resolved XPS measurements) and the water contact angle. Priest and co-workers have exploited ToF-SIMS as a possible predictive tool for surface wettability [27]. Patterned hydrophilic and hydrophobic regions self-assembled monolayers surfaces were correlated with ToF-SIMS fragmentation of the hydrophobic and hydrophilic surface groups and static contact angle measurements. A clear relationship between the surface wettability and relative intensity of characteristic secondary ions was identified. They claimed that their results for planar surfaces can be extrapolated to predict the wettability of particulate samples for which direct contact angle measurements are not straightforward. Surface hydrophilicity resulting from the co-effect due to incorporation of oxygen functional groups and creation of charge states is investigated by Chai *et al* [28].

Ruiz *et al* also demonstrated a clear link between the chemistry of the etched photoresist surface and hydrophobicity. C-F groups detected on the surfaces were thought to be responsible for the increase in hydrophobicity after the photoresist was etched [29]. All of these studies show that the wettability has a strong correlation with the chemistry of the surface determined by surface analytical techniques.

3. ToF-SIMS Spectra in the Application of Predicting the Wettability

To demonstrate the concept, a method for the prediction of wettability of patterned ITO surface is discussed here [16]. This is achieved through the construction of a PLS model using ToF-SIMS spectra and WCA data from a series of differently treated ITO surfaces. The model displays a reasonable correlation of the WCA values to the SIMS data. To test the predictive ability of the model, ITO surfaces patterned with hydrophilic and hydrophobic regions were examined using ToF-SIMS imaging to produce maps of secondary ion intensity. Our preliminary results show that the model can predict the spatial variations in WCA across the sample using ToF-SIMS data only. Such methodologies, if developed further, may be able to obtain quantitative predictions of surface wettability.

3.1. Microscale Wettability Mapping

ITO layers (45 nm thickness, ~ 12 mm × 12 mm size) deposited on glass plate (supplied by Cambridge Display Technology, UK) were used as substrates. A series of cleaning treatments including solvent cleaning, Ultraviolet Ozone (UVO) cleaning or a combination of both to provide a wettability variation across the surface. Four regions of each sample were separately analysed by SIMS to correlate with regional WCA measurements (see Figure 5a). The treatments were chosen to cover a broad range of contact angles. This dataset was used to construct the PLS model and is thus termed the calibration set.

For testing the predictive ability of the PLS model, a patterned ITO surface modification was performed to obtain the dataset (termed the test set) using SIMS imaging data. A mask with a hole (6 mm in diameter) was placed over the ITO surface with a small gap between the mask and the ITO. The sample was then UVO treated. This modifies the wettability of the region under the hole, which was directly exposed to UV light.

Contact angle measurements were carried out on all the samples immediately after surface treatment. After SIMS analysis, the samples were re-analysed by contact angle measurement to check for any changes. A water droplet with a base diameter of ~ 120 μm was dispensed by a piezo-dosing unit onto the ITO surface using a DSA100 (Krüss, Germany) drop shape analyser in the sessile drop configuration. Ultrapure water was used for all WCA measurements. For the ITO calibration set, a square array of measurement points covering 6 mm × 6 mm at 1.5 mm intervals (in total 16 measurements) was used (see Figure 5a). The mean WCA for 4 data points was used to determine the contact angle for each region of the sample. To map the wettability of the patterned ITO sample, a square array of 9 mm × 9 mm at 1.2 mm intervals (in total 64 measurements) were used (see Figure 5b). Due to water droplets evaporating in seconds, videos of the droplet profile were recorded. The first frame of the stable droplet landing onto the surface was chosen to determine the WCA by using a circle-fit function to the drop profile.

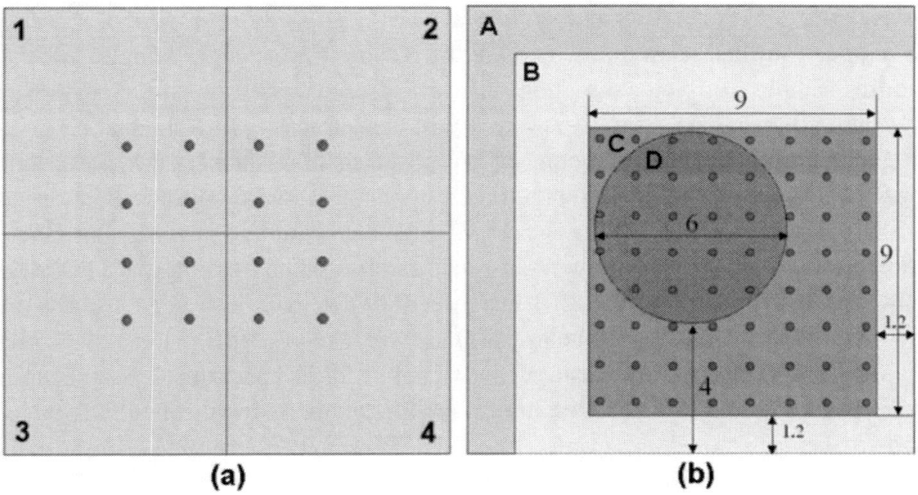

Figure 5. (a) A schematic diagram of ITO contact angle map for the PLS model. (b) A schematic diagram showing the geometry of the patterned ITO contact angle map: A is the mask, B is the ITO sample, C is the contact angle map area, D is the area exposed to UVO treatment. The dots on the contact angle map area show the position of the droplets deposited. The numbers in the figure are in millimetres. Reprinted with permission from ref. [16]. Copyright © 2010 John Wiley & Sons, Ltd.

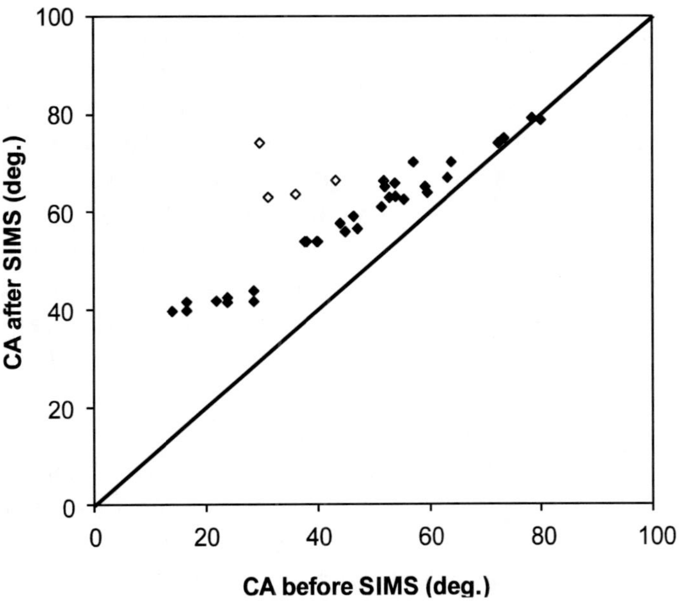

Figure 6. Measure WCA before SIMS versus measured WCA after SIMS for ITO calibration dataset (◆). Unfilled markers (◇) in the figure show the outliers from one single sample, which had been excluded from the calibration set. The line in the figure represents no change in contact angle. Reprinted with permission from ref. [16]. Copyright © 2010 John Wiley & Sons, Ltd.

Figure 6 compares the WCA before and after SIMS analysis. It was found that the WCA of ITO surfaces within the calibration set increased after SIMS analysis. For WCA above 70º, little increase is observed. For WCA between 40-70º ranges, the measured WCA after SIMS increased by 10-15 degrees, while if the measured WCA value before SIMS is lower than 40º, the increase after SIMS measurement is more than 15 degrees. From Figure 6, it seems there are four outliers (unfilled markers), which are the results from a single sample. We therefore excluded this sample from the calibration set before PLS analysis. In total, 32 SIMS spectra and WCA measurements on 8 samples were used to construct the PLS model. To investigate the cause of the WCA increase, ageing of UVO treated ITO samples has been performed either in fluoroware wafer trays (Entegris, Inc) in ambient laboratory atmosphere or mounted in the SIMS sample holder in the vacuum chamber without carrying out any SIMS acquisition. Only a slight increase in WCA value is observed by storing the ITO in a wafer box for 4 hours. While more than 10º increase of WCA has been seen for 4 h vacuum exposure of an identical UVO treatment of ITO surface. We note that the vacuum stored ITO also demonstrated higher organic contaminant signal in SIMS spectrum. It has been reported that surfaces exposed to vacuum conditions are prone to organic contamination pick-up compared to surfaces stored in a wafer box [30]. Therefore, the primary cause of contact angle increase in our study was exposure to the vacuum environment of the SIMS chamber. A quick effective SIMS measurement is essential to minimise changes in the surface chemistry and accumulated contamination within the SIMS instrument.

3.2. PLS Model in Related to Wettability Prediction

To build a PLS model, PLS_Toolbox 5.2 (Eigenvector Research, Inc., WA) in conjunction with Matlab version 7.6.0.324/R2008a (Mathworks, Inc., MA) was used. SIMS spectra and images were taken using a ToF-SIMS IV (IONTOF GmbH, Münster, Germany) time-of-flight secondary ion mass spectrometer. Care was taken to make sure the analysis area was aligned to the area where water droplets landed for WCA measurement. For the calibration set, a positive spectrum was acquired from each of the four regions on each sample. The SIMS data were normalized to the total ion count (between 1 u to 200 u, excluding indium 115 u), to account for the variation in the total secondary ion intensity between samples. For the acquisition of SIMS images on the patterned ITO sample (the test set), the ion beam was rastered over 64 × 64 pixels and positive mass spectra acquired at each pixel, typically 64 shots per position were used for each image. A 9 mm × 9 mm square area of positive ion images was achieved by additionally rastering the motorized stage. The normalized secondary ion intensities and contact angle data were mean-centred prior to the subsequent analyses. The peak due to the main isotope of indium (mass 115 u) was removed from the peak list due to the peak saturating the detector in many datasets. The minor isotope of indium at mass 113 u was included which did not exhibit this problem. The Root Mean Square Error of Calibration (RMSEC) was calculated to quantify how well each model fitted the calibration set. The Root Mean Square Error of Prediction (RMSEP) was used to determine the predictive ability of the models to the test set.

We find a better correlation between SIMS and WCA (i.e. a lower RMSEC) when the WCA is measured after SIMS analysis, probably because the WCA measured after SIMS may reflect the actual surface chemical properties of the ITO surface after exposure to the vacuum environment during the SIMS acquisition. When the experimental values of WCA are plotted against those predicted by the PLS models (Figure 7a), a linear relationship with a relatively low RMSEC is achieved, indicating a reasonable fit of the PLS model to the calibration set. In the regression coefficient plot Figure 7b, the ions with the largest positive regression coefficient and associated with high contact angles have masses of 41 ($C_3H_5^+$), 57 ($C_4H_9^+$), 77 ($C_6H_5^+$), 91 ($C_7H_7^+$), 107 ($C_8H_{11}^+$) and 129 ($C_{10}H_9^+$). The peaks with largest negative coefficients are strongly associated with low contact angle; they have masses of 15 (CH_3^+), 27 ($C_2H_3^+$), 46 ($C_2H_6O^+$), 113 (In^+), 132 ($InOH^+$) and 148 (InO_2H^+). The results are sensible, in that hydrocarbon $C_mH_n^+$ fragments generally correlate with high contact angles due to the hydrophobicity of these groups, while secondary ion fragments containing Indium indicate exposure of the hydrophilic oxide substrate and correlate with lower contact angles. The results indicate that the SIMS spectra do correlate strongly with the wettability of the surface.

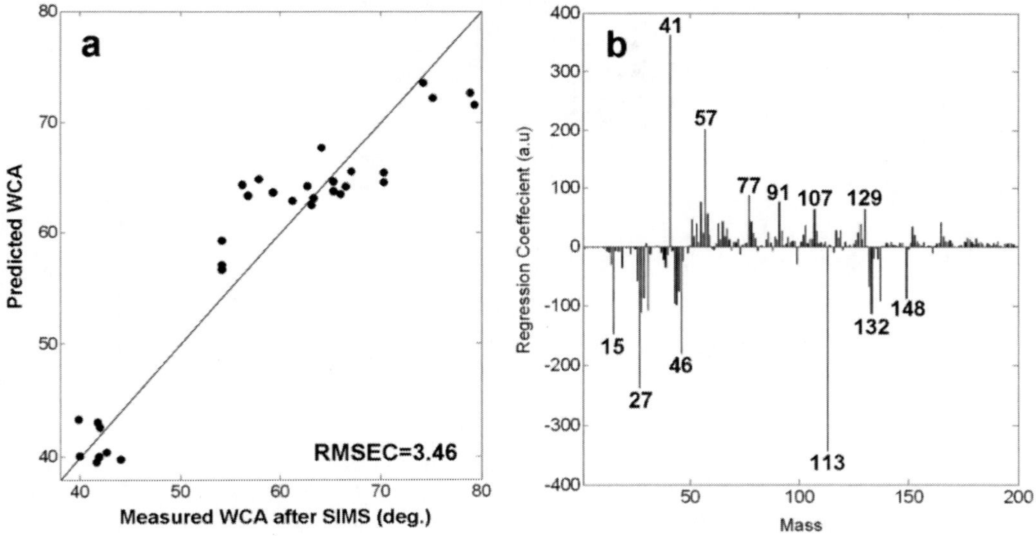

Figure 7. PLS model constructed using the SIMS and the WCA data from the calibration set. Both data were mean centred: (a) Measured WCA for ITO calibration set versus predicted WCA values. The line represents equality between the measured and predicted WCA. (b) Regression coefficient versus mass. Reprinted with permission from ref. [16]. Copyright © 2010 John Wiley & Sons, Ltd.

In order to test the predictive ability of the PLS model, patterned ITO samples with different UVO treatment times were analyzed by WCA analysis and SIMS imaging to form a test set. The ToF-SIMS image data was mean-centered using the mean from the calibration set. The PLS models, obtained using the calibration data above, are used to predict the WCA using SIMS image data only. These predicted values are then compared to independently measured experimental WCA data. An example of an ITO patterned sample is shown in Figure 8a. Sodium ion SIMS images (not shown here) from patterned ITO samples show a clear 8 by 8 grid array where the water droplets evaporated from the substrate surface and deposited trace amounts of sodium. Figure 8b shows the predicted WCA maps generated from the 9 mm × 9 mm SIMS image, using the models calibrated previously with the WCA measured before SIMS. Although the predicted WCA have some errors (RMSEP = 10.5°), the spatial correlation between the high and low contact angle regions is good. The area that was exposed to UVO modification has lower contact angles as expected by removal or oxidation of hydrocarbon contaminants. The predicted WCA is plotted against the actual WCA values (Figure 8c) and shows correlation between the values, but a rather high level of scatter. Such discrepancies may possibly be due to the pre-processing of the data prior to analysis, some spatial offset between the SIMS and contact angle images, the presence of contaminants or other contributions to peak intensities that were not within the calibration set, or the low signal-to-noise ratio of the spectra derived from the SIMS image. It is expected that the surface wettability predictions may be improved by more sophisticated data pre-processing. The correlation between SIMS and WCA after SIMS analysis is much better, which is as expected since after exposing to the vacuum environment SIMS can reflect the actual surface (not shown here, see ref. [16] figure 4).

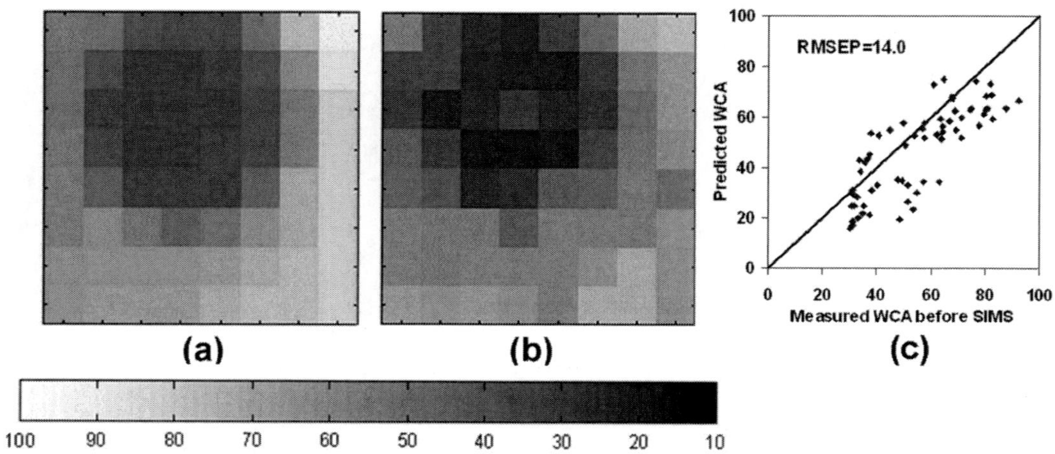

Figure 8. Results of PLS predictions obtained on a test data set consisting of an ITO sample patterned by 2 minutes UVO treatment. (a) Measured contact angle map before SIMS. (b) Predicted contact angle map from the SIMS data, using the model calibrated with WCA measured before SIMS. (c) Plot of predicted WCA versus measured WCA, for WCA measured before SIMS. Reprinted with permission from ref. [16]. Copyright © 2010 John Wiley & Sons, Ltd.

4. Correlations between XPS Composition and Wetting Properties

We have demonstrated that very clear, linear correlations between XPS spectroscopic data and contact angles may be obtained using this methodology in another example [17].Using plasma-treated organic light-emitting-diode (OLED) substrates of indium-tin-oxide (ITO) and ZPN1168-30 resist with chemical formula $C_6H_mOH(CH_3)_{4-m}(CH_2)_n$ (ZPN). XPS gives precise surface chemical information, which can be a powerful indicator of surface wettability. The prediction for ZPN regions is slightly better than for ITO regions, their standard uncertainties in the cosine of the wetting angle being 0.07 and 0.10, respectively. The bonding groups were identified that control the water wettability. The high-energy resolution of C 1s data also provided insights into the quantitative assessment of the surface functional groups, allowing the prediction of the surface wettability for industries where direct contact angle measurement is not possible.

4.1. Experimental Details

OLED substrates (0.07 cm thick, 25 mm × 25 mm size) consisted of glass plates with 140 nm plasma-sputtered ITO. This is used for the OLED anode and is called, here, the "ITO region". The ITO surface was approximately half-patterned with 1.4 μm thick ZPN (from Zeon Corp, Tokyo) using standard optical lithography. This resist is for the bank material shown in Figure 1 of reference [31] and is called, here, the "ZPN region". A plasma reactor was used to modify these OLED substrates by oxygen (O_2, labeled as O_2 treatment), tetrafluoromethane (CF_4, labeled as CF_4 treatment), or O_2 followed by 60 second CF_4 plasma (labeled as O_2/60 s CF_4 treatment). The treatment conditions (16 samples in total) were selected according to previous studies [32] and were maintained constant, i.e. an inter-electrode gap of 5 cm, a gas pressure of 30 mTorr, a gas flow rate of 30 standard cubic centimeter per minute (sccm), and an RF-power ~ 1 W. Treatments were carried out as a function of time between 0 and 300 s. More details of the plasma setup can be found in ref [31].

The contact angles of plasma-treated samples were measured approximately 15 minutes after the end of plasma treatment using a CAM100 (KSV Ltd.) contact angle meter in a laboratory ambient of 18 ± 2 ºC. Ultrapure de-ionized water was used for all the contact angle measurements. Three to five drops of 1.2 μl were dispensed along each region (i.e. ITO and ZPN) and ten frames (1/second) were captured after depositing each drop. The interval between the drop landing and the camera triggering was 3 s to allow time for establishing thermodynamic equilibrium of the liquid on the solid surface. The standard deviation of the measurements in these successive frames was typically 3° for ITO and 1° for ZPN. These measurements, rather than an average contact angle for, say, the advancing and retreating contact lines, provide a simulation of the ink jet drops hitting these surfaces in the printing of OLED devices. No hysteresis measurements were made in this study. The Young-Laplace theory [19] was used in the image-processing software to determine the contact angle from the drop profile on the substrate.

XPS measurements were conducted with an AXIS Ultra X-ray spectrometer (Kratos Analytical Ltd, Manchester, UK) with monochromatic Al Kα radiation (hν= 1486.6 eV) at normal emission. Three separate areas of approximately 700 μm × 300 μm on each ITO region and each ZPN region were analyzed on each sample. Charge compensation, using low energy electrons to flood the surface, was applied during all measurements. The acquired data were processed for both quantification and peak fitting using Casa XPS 2314 software. Surface elemental compositions were calculated from the peak areas using a linear background. The C 1s peaks were fitted by Gaussian (70%) - Lorentzian (30%) functions with a Shirley background to measure each chemical state. For the ZPN region, an energy calibration of 285 eV was applied to the main C 1s peak for all samples.

4.2. XPS Quantification

It was observed in Figure 9 that the contact angle of both regions on the OLED patterned substrates decreased after O_2 treatment. The contact angle falls sharply after a short treatment time and then approaches a value around 15° after a longer treatment time (300 s) on both regions. For CF_4 treatment, a great contrast between WCA on the two regions is achieved after a long exposure to the plasma. The ZPN hydrophobicity smoothly increases from the start of the treatment and remains relatively constant for longer treatment times. The water contact angle on ITO rises to 99° in the first 10 s and then progressively decreases to 25° after 300 s. Similar experiments were carried out to study the effect of a two-step treatment (O_2/CF_4 treatment: various length O_2 plasma treatments followed by 60 s of CF_4 plasma) on each region. Only 30 s O_2/60 s CF_4 treatment provides an ITO contact angle ~ 20° and a ZPN contact angle ~ 111°. By varying the processing time with the two plasmas, a high flexibility for controlling the wettability on the two regions is achieved.

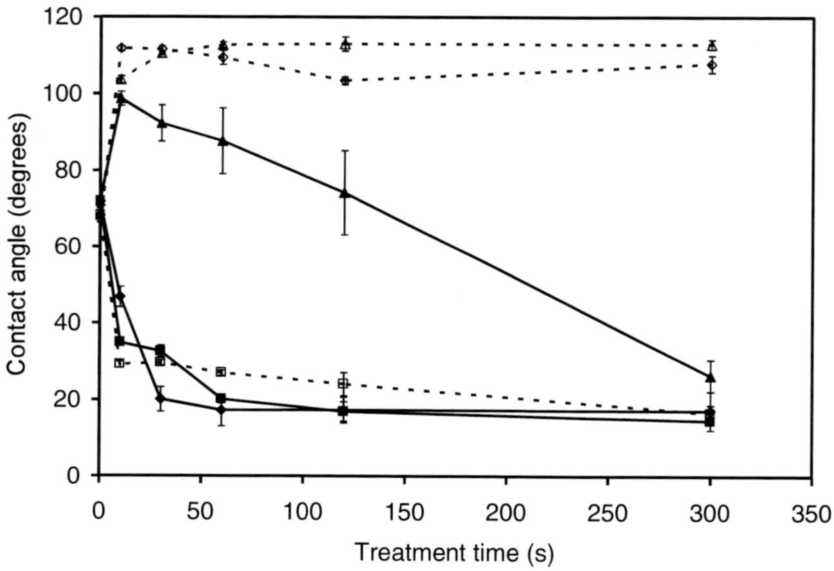

Figure 9. The contact angle for different plasma treated times, (■) ITO O_2 treatment, (▲) ITO CF_4 treatment, (♦) ITO O_2/60 s CF_4 treatment, (□) ZPN O_2 treatment, (Δ) ZPN CF_4 treatment and (◊) ZPN O_2/60 s CF_4 treatment. Reprinted with permission from ref. [17]. Copyright © 2010 American Institute of Physics.

Table 1. Peaks used for C 1s peak fitting in XPS spectra

Identifier	C1	C2	C3	C4	C5	C6
Peak centre, eV	285.0	286.6	288.0	289.3	291.6	293.7
Chemical state	C-C	C-O-C	C=O	O-C=O/CF	CF_2	CF_3
Beamson and Brigg's peak centre, eV. CF_x values are taken from Viton A.	285.0	286.6	287.9	289.3/289.8	291.5	293.9

Reprinted with permission from ref. [17]. Copyright © 2010 American Institute of Physics.

In order to investigate the causes of the contact angle changes, the surfaces were characterized by XPS to explore the detailed surface chemistry after different plasma modifications. Typical spectra for the ITO and ZPN substrates before and after treatment are shown in Figs. 10 and 11, respectively. The peaks and their assignations are given in Table 1. In the fitting, after referencing the binding energies to the C1 peak at 285.0 eV, the peak centers were fixed at the values tabled for all spectra by an iterative process. The peak widths were all fixed to be the same as that for the main carbon peak.

For ITO patterns, no obvious change was observed for the sample after 300 s O_2 plasma treatment compared to the untreated ITO surface. Peak synthesis for the C 1s spectrum gave four main peaks, which are assigned to C-H (C1: 285.0 eV), C-O-C (C2: 286.6 eV), C=O (C3: 288.0 eV) and O-C=O (C4: 289.3 eV) as shown in Figure 10c and Table 1. The C-H could also be C-C and the C-O-C could also be C-OH. These values are given by Beamson and Briggs [33] as shown in Table 1 but with scatters of, generally, 0.1 eV. When the samples were treated with the CF_4 plasma, two further peaks appeared at a higher binding energy of 291.6 eV (C5) and 293.7 eV (C6), indicating the fluorinated carbon species as shown in Figure 10c. This is close to the CF_2 value of either 291.2 or 291.8 eV and the CF_3 value at 293.9 eV given by Beamson and Briggs for Viton A. A similar spectrum was found for 300 s O_2/ 60 s CF_4 treated ITO sample but without the C6 peak, as shown in Figure 10d.

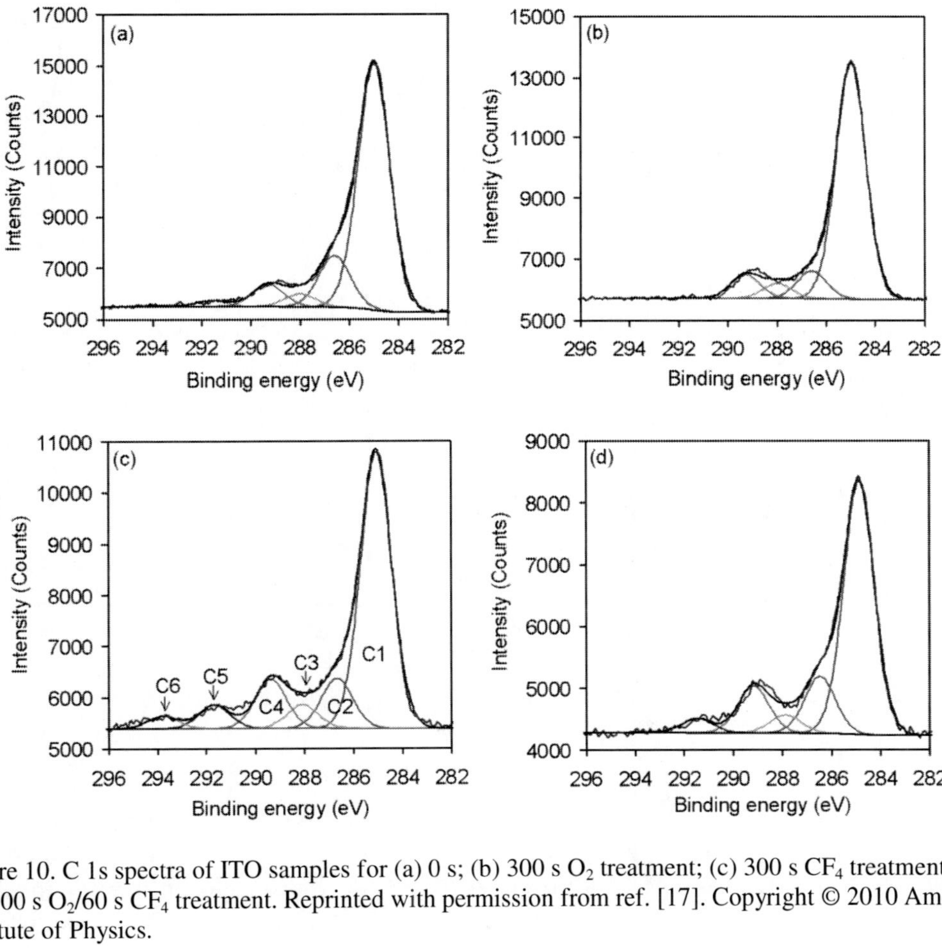

Figure 10. C 1s spectra of ITO samples for (a) 0 s; (b) 300 s O_2 treatment; (c) 300 s CF_4 treatment and (d) 300 s O_2/60 s CF_4 treatment. Reprinted with permission from ref. [17]. Copyright © 2010 American Institute of Physics.

The C 1s spectrum of the virgin ZPN pattern exhibited four principal bonding environments as shown in Figure 11a. A strong peak resulting from the hydrocarbon component (C1) and was used as a reference for calibration. Components were additionally observed that were attributed to ether (C2) and ester (C3) environments as above (labels shown in Figure 11c). A small shake-up satellite was seen at ~ 291.3 eV. Following O_2 plasma exposure as shown in Figure 11b, the ester peak broadened and the hydrocarbon, ether, and ester peaks were present in a 8.8:2.2:1 ratio, which is much lower than the 20.2:4.5:1 ratio of virgin ZPN. This is in agreement with a loss in carbon content by partial removal of the surface contaminated layer, which then resulted in a low contact angle. Figure 11c shows the C 1s spectrum acquired after 300 s of the CF_4 plasma treatment. The most significant change in the spectrum was the appearance of a new peak at 293.7 eV (C6), corresponding to CF_3 bonding given by Beamson and Briggs at 293.9 eV for Viton A [33].

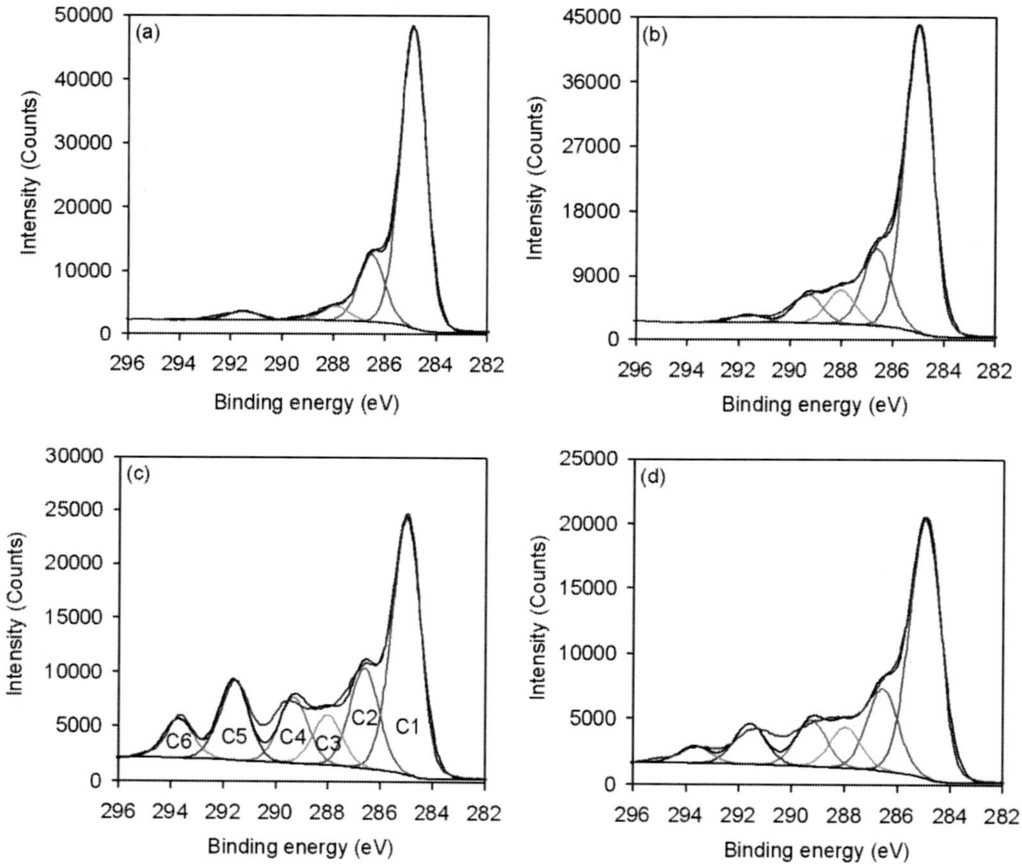

Figure 11. C 1s spectra of ZPN samples for (a) 0 s; (b) 300 s O_2 treatment; (c) 300 s CF_4 treatment and (d) 300 s O_2/60 s CF4 treatment. Reprinted with permission from ref. [17]. Copyright © 2010 American Institute of Physics.

A substantial increase was observed in the size of the C5 component. Such increase and the appearance of the new component were accompanied by a reduction in the amount of C1 carbon from 58.15% (0 s) to 33.14% (300 s CF_4 treatment). Note also the increase in intensity and narrowing of the C4 peak which could be O-C=O as described above but which is also probably unresolved CF at 289.8 eV (for Viton A). The fall in the fraction of C1 carbon was found to be closely correlated with the fall in the total percentage of surface carbon. It is interesting to note that the intensities of C5 and C6 components are much lower for 300 s O_2/60 s CF_4 treated ZPN sample as shown in Figure 11d. The spectrum also exhibited a notable change at higher binding energies for 300s O_2/ 60 s CF_4 exposure compared to sample treated by 300 s O_2 treatment as shown in Figure 11b.

4.3. Simple Model for Correlating XPS Spectra with Wettability

A typical XPS measurement gives information about the surface with a characteristic depth of 3 nm but a total sampling depth of ~ 10 nm, whereas the contact angle value is sensitive to the topmost few angstroms due to the nature of the forces involved in the wetting process [34]. The sampling depth of the water is thus about 0.5 nm [35]. Such extreme surface sensitivity must be kept in mind when comparing wettability data and results obtained with other surface sensitive techniques, such as XPS in this study. Only the major elements (such as C, O and F) present in the plasma-treated surfaces will be included from the following correlation since they play important roles to the wetting characteristics of the surface.

As discussed in section 2, if one can define γ_{SG} and γ_{SL}, the contact angle can be defined according to Young's Eq. (1). XPS is a direct measurement of the chemical environment of the atoms at the surface. The components on the topmost surface govern the total surface free energy in the materials. Therefore, we may apply the XPS data illustrated in Figs. 10 and 11 to correlate with the WCA data given in Figure 9. Due to the two substrates are rather different in chemistry, separate models have been generated to provide a better correlation. More details for the model can be found in ref. [17].

We first consider the ZPN data. Here we have data for the C, O and F atomic fractions within the analysis depth of XPS, together with the C split into C-H, C-O-C, C=O, O-C=O/CF, CF_2 and CF_3 states. Correlations show that, approximately $X_{CF} + 2X_{CF2} + 3X_{CF3} = X_F$ except at very low X_F where the C4 peak is for the O-C=O state but at higher X_F values it is for CF. Either the sum of the CF_n fractions or the F fraction may be used to identify the effect of the hydrophobic carbon bonded to fluorine contribution at the surface. The similar analysis with $0.5X_{C\text{-}O\text{-}C} + X_{C=O}$ shows this to be significantly higher than X_O and so must include other states. Observation of the correlation of the composition with $\cos\theta$ shows that, in general, as $\cos\theta$ falls (i.e. as wetting declines) the value of X_F from the F 1s intensity rises. In the same data set $X_{C\text{-}H}$ falls strongly and $X_{C\text{-}O\text{-}C}$ falls weakly whilst $X_{C=O}$ is roughly constant. Both $X_{C\text{-}O\text{-}C}$ and $X_{C=O}$ are significantly less than $X_{C\text{-}H}$ and do not change sufficiently here to allow the behavior to be evaluated. That $X_{C\text{-}H}$ falls whilst X_F rises in this series does not show that they have opposite effects energetically but simply that the treatment enhancing the F content also reduces the C-H content. Figure 12a shows how the significant XPS compositions change with the three treatments and as a function of time. Using a least squares fitting, we find that $\cos\theta$ is largely affected by two compositions

$$\cos\theta = B + b_F X_F + b_{C\text{-}H} X_{C\text{-}H} \tag{2}$$

as shown in Figure 12b with B as a constant here. This correlation is very good and shows, with $B = 3.44$, $b_F = -0.078$, $b_{C\text{-}H} = -0.050$, an excellent standard uncertainty of 0.07 in using Eq. (2) to predict $\cos\theta$. The error bars show the 95% confidence limits for the deviations of the means of the experimental values of $\cos\theta$, $\cos\theta_{expt}$, rather than the standard deviations used earlier.

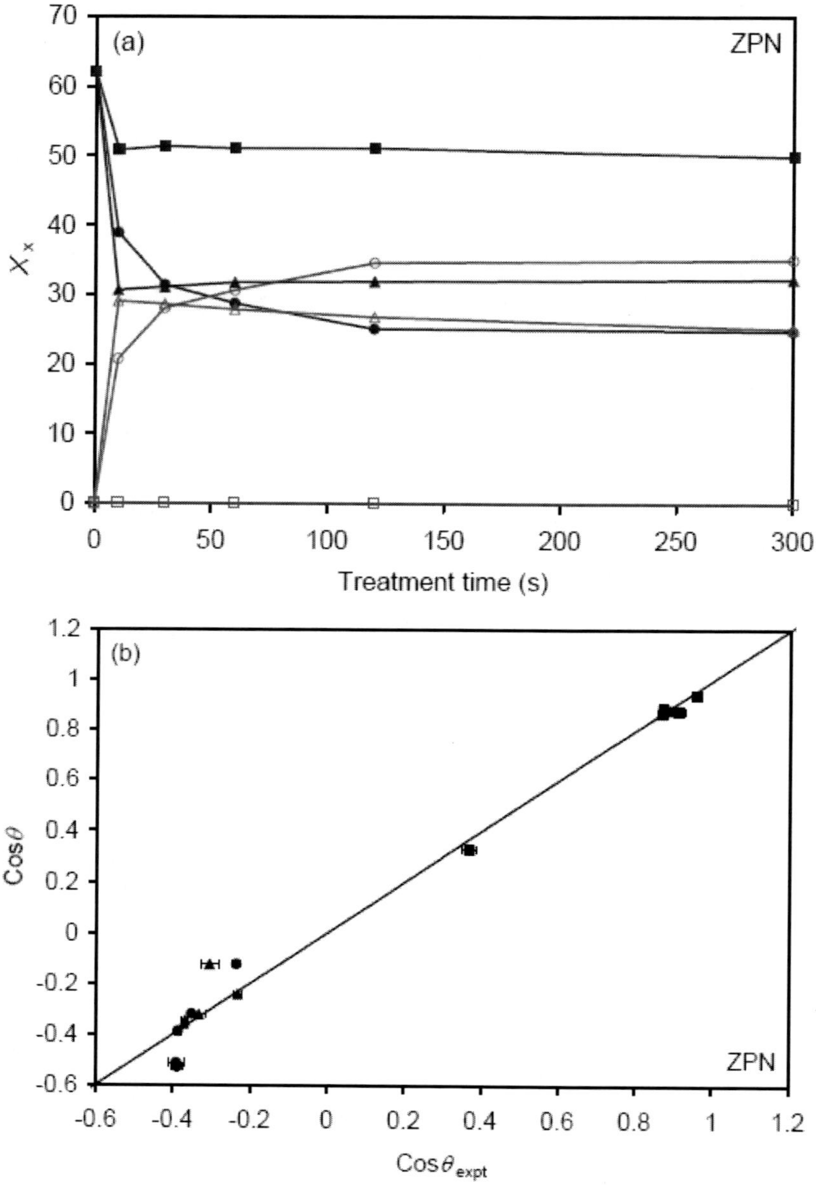

Figure 12. XPS data and the treatment of ZPN. (a) The evolution of different compositions with treatment time. The square, circle and triangle markers are assigned to O_2, CF_4 and O_2/60 s CF_4 plasma treatment, respectively. The solid markers denote the atomic fraction of the C-H group and open markers the fluorine atomic fractions. Key: C-H (solid marker), F (open maker). (b) Correlation between the measured values of $\cos\theta$ and the fitted relations from XPS with plasma treatments: (■) O_2, (●) CF_4, (▲) O_2/60 s CF_4. Reprinted with permission from ref. [17]. Copyright © 2010 American Institute of Physics.

The XPS results for ZPN in Figure 12a show that very rapidly, for the CF_4 and O_2/60 s CF_4 treatments, the hydrophobic saturated carbon is replaced by the even more hydrophobic fluorinated carbon such that, as shown in Figure 12b, the strongly dewetting contact angle of ~ 110º is reached. On the contrary, without the CF_4, the saturated hydrocarbon falls less but there is no counteracting fluorinated carbon to increase the wetting angle. With both O_2 treatments, there is a small increase in X_O but this cannot be separated, with the present data, from the concomitant falls in X_{CF} or $X_{C\text{-}H}$.

The situation with ITO is rather different. According to the wide scan spectra (not shown here), it clearly shows a thin or very thin and patchy overlayer on top of the ITO substrate. The correlations of the In and Sn fractions with X_O show $X_O/(X_{Sn} + X_{In}) = 1.98$ with a scatter of ~ 5%.

This indicates that the O 1s peak observed is mainly from the substrate and that, if there is any extra surface O, it will give a signal that may be swamped by the substrate signal. Observation of the correlation of the composition with $\cos\theta$ shows that, in general, as $\cos\theta$ falls, X_F rises but $X_{C\text{-}H}$ is approximately constant and $X_{C\text{-}O\text{-}C}$ and $X_{C=O}$ are both small. As X_F rises, X_O falls although, as noted above, that change occurs as the substrate is being covered. The anti-correlation of X_F and X_O means that, as discussed earlier, their effects cannot be separated.

Figure 13a shows how the significant XPS compositions change with the three treatments and as a function of time. Using a least squares fitting, we find that $\cos\theta$ may be described by

$$\cos\theta = B + b_O X_O + b_F X_F + b_{C\text{-}H} X_{C\text{-}H} + b_{C\text{-}O} X_{C\text{-}O} \tag{3}$$

as shown in Fig 13b. Again, the error bars show the 95% confidence limits for the deviations of the means of $\cos\theta_{expt}$, rather than the standard deviations used earlier. This correlation is very good and shows, with $B = 0.515$, $b_O = 0.019$, $b_F = -0.019$, $b_{C\text{-}H} = -0.012$, and $b_{C\text{-}O\text{-}C} = -0.136$, a standard uncertainty in using Eq. (3) to predict $\cos\theta$, an excellent 0.10. In defining these parameters, b_O was set to be $-b_F$ since their anticorrelation meant that these parameters were related in the minimization. Additionally, $b_F/b_{C\text{-}H}$ was set as for ZPN. A free choice of parameters led to a slight further reduction in uncertainty.

The XPS results in Figure 12a show smaller changes than in Figure 13a but that, very rapidly, for the CF_4 and O_2/60 s CF_4 treatments, the hydrophobic saturated carbon is replaced by the even more hydrophobic fluorinated carbon such that, as shown in Figure 6, the strongly dewetting contact angle of ~ 100º is rapidly reached in the case without O_2. This is mainly caused by the sudden rise in the F fraction causing a fall in O, In and Sn fractions followed by a slow reversal of this behaviour so that the wetting significantly improves.

These correlations show that XPS is very powerful for relating the surface compositions to materials properties and is good for interpolating properties for the studied systems. Extrapolating outside the range or to other materials is, however, unlikely to succeed without significantly more data. These data sets are very limited. In many cases, as one constituent rises, so does another so that one could be tempted to conclude that if both were rising with $\cos\theta$, then both contributed to better wetting.

However, one could be working against the other and separating their effects, as noted earlier, may be difficult. The only way to separate the components properly is with data sets where single constituents are separately changed over a significant range. In practice, a pair must be altered since one bond is replaced by another.

In the practical problems studied here, although many of the changes are linked, clear conclusions may be drawn about the changes that improve or degrade the wetting and about the wetting characteristic of analyzed surfaces.

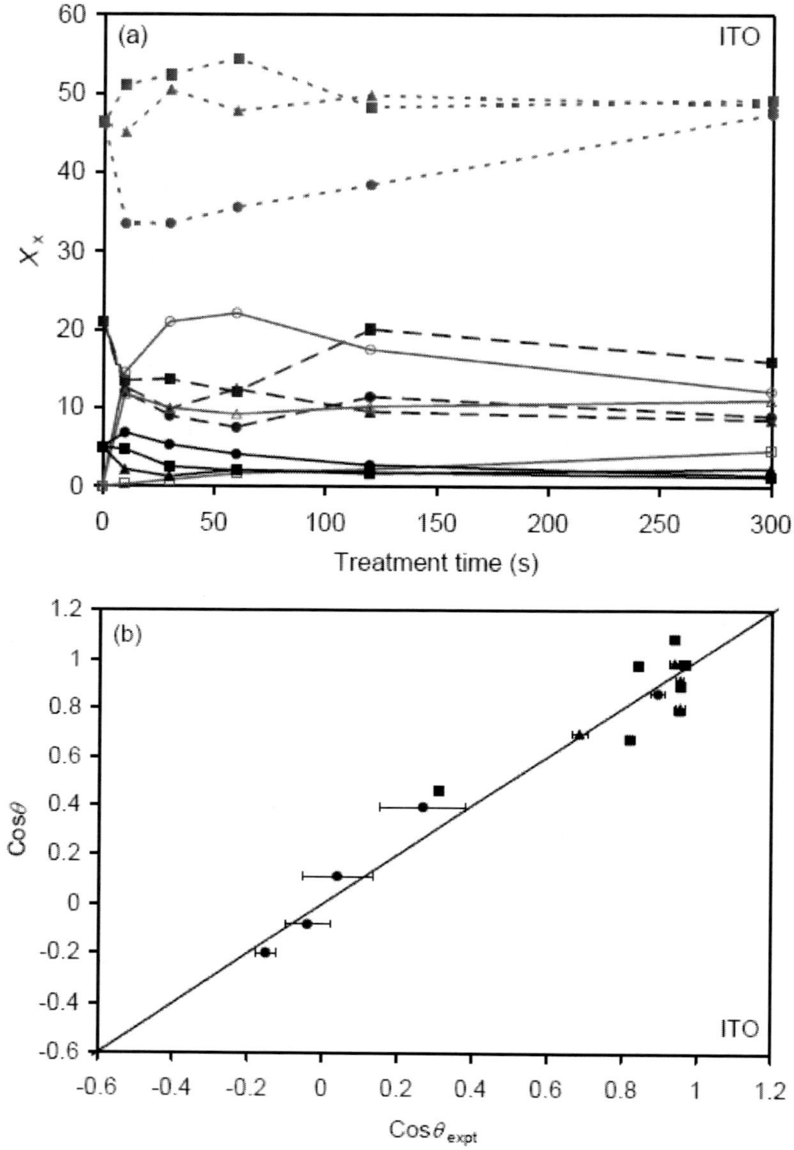

Figure 13. XPS data and the treatment of ITO. (a) The evolution of different compositions with treatment time. The square, circle and triangle markers are assigned to O_2, CF_4 and O_2/60 s CF_4 plasma treatment, respectively. Key: C-O-C (solid marker with solid line), O (solid marker with short dash line), C-H (solid marker with long dash line), F (open maker with solid line). (b) Correlation between the measured value of cosθ and the fitted relations from XPS with plasma treatments: (■) O_2, (•) CF_4, (▲) O_2/60s CF_4. Reprinted with permission from ref. [17]. Copyright © 2010 American Institute of Physics.

CONCLUSION

In this chapter, we have shown that surface analysis techniques such as ToF-SIMS and XPS are powerful for identification, quantification and data obtained from spectra or images can be correlated with surface chemical characteristics such as wettability. In contrast with conventional direct contact angle measurement, this strategy can be an alternative method for the prediction of the surface wettability for the industries (such as microfluidics, plastic electronics and geometry-constrained devices) where direct liquid droplets cannot be placed in the interested surfaces or size-constrained features. Two examples have been given to demonstrate this concept. Earlier work [16] in differently treated Indium Tin Oxide (ITO) surfaces system, using partial least squares analysis with SIMS with detailed spectra covering many significant spectral peaks showed that non-oxygenated hydrocarbon fragments $C_nH_m^+$ are correlated with higher WCA values, while oxygenated fragments $C_nH_mO^+$ and indium fragments are correlated with low WCA values. This agrees with the common understanding that nonpolar species are hydrophobic and polar species are hydrophilic, and demonstrates the potential of using ToF-SIMS for building models between surface chemistry and resulting wettability properties. A further example [17] is given of XPS spectroscopic data obtained from the plasma-treated OLED substrates of the ITO and the photoresist surfaces (ZPN) to evaluate the contact angle in order that the wettability of very small areas may be predicted. Very clear correlations for both the ITO and ZPN are obtained enabling the prediction of the contact angles for these plasma-treated OLED materials from XPS data to a standard uncertainty of 9% in $\cos\theta$. The use of static SIMS instead of XPS has the major advantage of providing information over the required outermost atom layer. However, the lack of quantitative signals in SIMS makes the data less tractable than for XPS. For SIMS image maps where the signal levels were low, this led to correlations of the predicted and measured contact angles that were significantly poorer than the correlations obtained by XPS work. It is advised to choose the appropriate surface analysis technique in order to get a good correlation with wetting properties. These correlations also need to be extended to a wider range of compositions in order to establish the physicochemical properties of particular surface functional groups that control water contact angles by this method.

ACKNOWLEDGMENTS

The authors would like to thank J. L. S. Lee for assistance with the PLS model and P Svarnas and J W Bradley from University of Liverpool for plasma treatment on OLED samples and WCA measurement. We also thank M P Seah for extensive discussions on the issues covered in the present chapter. This work forms part of Technology Strategy Board (Project: Surface Conditioning of Plastic Electronics No. TPM1012C) and the Chemical and Biological Metrology Programme of the National Measurement System of the UK Department of Business, Innovation and Skills (BIS).

REFERENCES

[1] Terry, S. C.; Jerman, J. H.; Angell, J. B. *IEEE Trans. Electron.Devices.* 1979, 26, 1880-1886.
[2] Bassous, E.; Taub, H. H.; Kuhn, L. *Appl. Phys. Lett.* 1977, 31, 135-137.
[3] Manz, A.; Graber, N.; Widmer, H. M. *Sens. Actuators* 1990, B1, 244-248.
[4] Verpoorte, E. *Electrophoresis* 2002, 23, 677-712.
[5] Telleman, P.; Larsen, U. D.; Philip, J.; Blankenstein, G.; Wolff, A. *Proceddings of Micro Total Analysis Systems 1998*; Kluwer Academic Publishers: Dordrecht, The Netherlands, 1998; pp 39-44.
[6] Woolley, A. T.; Mathies, R. A. *Anal. Chem.* 1995, 67, 3676-3680.
[7] Meinhart, C. D. *J Microelectromech. S.* 2000, 9, 67-75.
[8] Haswell, S. J.; Middleton, R. J.; O'Sullivan, B.; Skelton, V.; Watts, P.; Styring, P. *Chem. Commun.* 2001, 5, 391-398.
[9] Huang, L. R.; Tegenfeldt, J. O.; Kraeft, J. J.; Sturm, J. C.; Austin, R. H.; Cox, E. C. *Nature Biotechnol.* 2002, 20, 1048-1051.
[10] Whitesides, G. M. *Nature* 2006, 442, 368-373.
[11] Reyes, D. R.; Lossifidis, D.; Auroux, P. A.; Manz, A. *Anal. Chem.* 2002, 74, 2623-2636.
[12] Auroux, P. A.; Lossifidis, D.; Reyes, D. R.; Manz, A. *Anal. Chem.* 2002, 74, 2637-2652.
[13] Erickson, D.; Li, D. Q. *Anal.Chim.Acta* 2004, 507, 11-26
[14] Stone, H. A.; Stroock, A. D.; Ajdari, A. *Annu. Rev. Fluid Mech.* 2004, 36, 381-411.
[15] Squires, T. M.; Quake, S. R. *Rev. Mod. Phys.* 2005, 77, 977-1026.
[16] Yang, L.; Shard, A. G.; Lee, J. L. S.; Ray, S. *Surf. Interface Anal.* 2010, 42, 911-915.
[17] Yang, L.; Svarnas, P.; Shard, A. G.; Bradley, J. W.; Seah, M. P. *J Appl. Phys.* 2010, 108, 114901-1-114901-8.
[18] Ismagilov R. *Angew. Chem. Int. Ed.* 2003, 42, 4130-4132.
[19] Young, T. *Phil. Trans. Roy. Soc. London* 1805, 95, 65-87.
[20] Baytekin, H. T.;Wirth, T.; Gross, T.; Treu, D.; Sahre, M.; Theisen, J.; Schmidt, M.; Unger, W. E. S. *Surf. Interface Anal.* 2008, 40, 358-363.
[21] Baytekin, H. T.; Wirth, T.; Gross, T.; Sahre, M.; Unger, W. E. S.; Theisen, J.; Schmidt, M. *Surf. Interface Anal.* 2010, 42, 1417-1431.
[22] Urquhart, A. J.; Taylor, M.; Anderson, D. G.; Langer, R.; Davies, M.; Alexander, M. R. *Anal. Chem.* 2008, 80, 135-142.
[23] Lee, J. L. S.; Gilmore, I. S. in *Surface Analysis: The principal techniques 2nd Edition*, Vickerman, J.C.; Gilmore, I. S.; Eds. Wiley, Chichester, 2009; pp 563-612.
[24] Taylor, M.; Urquhart, A. J.; Anderson, D. G.; Langer, R.; Davies, M.; Alexander, M. R. *Surf. Interface Anal.* 2009, 41, 127-135.
[25] von Gradowski, M.; Wahl, M.; Jacoby, B.; Hilgers, H.; Kopnarski, M. *Vakuum in Forschung und Praxis* 2005, 17, 73-79.
[26] Paynter, R. W. *Surf. Interface Anal.* 2002, 33, 14-22.
[27] Priest, C.; Stevens, N.; Sedev, R.; Skinner, W.; Ralston, J. *J Colloid Interface Sci.* 2008, 320, 563-568.
[28] Chai, J. N.; Lu, F. Z.; Li, B. M.; Kwok, D. Y. *Langmuir* 2004, 20, 10919-10927.

[29] Ruiz, A.; Valsesia, A.; Ceccone, G.; Gilliland, D.; Colpo, P.; Rossi, F. *Langmuir* 2007, 23, 12984-12989.

[30] De Gendt, S.; Knotter, D. M.; Kenis, K.; Depas, M.; Meuris, M.; Mertens, P. W.; Heyns, M. M. *Jpn. J. Appl Phys*. 1998, 37, 4649-4655.

[31] Svarnas, P.; Yang, L.; Munz, M.; Bradley, J. W.; Edwards, A.; Shard, A. G. *J. Appl. Phy*. 2010, 107, 103313-1-103313-10.

[32] Svarnas, P.; Bradley, J. W.; Shard, A. G. *Surf. Coat. Technol*. 2009, 204, 99-107.

[33] Beamson, G.; Briggs, D. *High Resolution XPS of Organic Polymers – The Scienta ESCA300 Database*, Wiley, Chichester, 1992; pp 1-295.

[34] Morra, M.; Occhiello, E.; Garbassi, F. *Adv. Colloid Interface* 1990, 32, 79-116.

[35] Defay, R.; Prigogine, I.; Bellemans, A.; Everett, D. H. *Surface Tension and Adsorption*, Longmans, London, 1966; pp1-432.

In: Microfluidics
Editors: Stefano Panzarella and Walter Maroni
ISBN: 978-1-62417-345-5

Chapter 5

INTEGRATION OF CAPACITIVE MICROMACHINED ULTRASOUND TRANSDUCERS TO MICROFLUIDIC DEVICES

Darius Viržonis,[1*] Galius Vanagas[1] and Rimantas Kodzius[2,3†]

[1]Kaunas University of Technology, Panevezys institute, Panevezys, Lithuania
[2]King Abdullah University of Science and Technology (KAUST), Computer, Electrical and Mathematical Sciences and Engineering Division (CEMSE), Thuwal, Saudi Arabia
[3]Envirola Project, Thuwal, Saudi Arabia

ABSTRACT

The design and manufacturing flexibility of capacitive micromachined ultrasound transducers (CMUT) makes them attractive option for integration with microfluidic devices both for sensing and fluid manipulation. CMUT concept is introduced here by presenting simplified equivalent circuit model; while finite element analysis (FEA) and near-field CMUT operation experiments were used as the main methods of research. The transient time FEA was used to simulate the transmission and reception of the short pulse and then analysis of the CMUT membrane motion in the time and frequency domains was made. 200 nm finite element spacing and 0.2 ns sampling time was used during the simulations for increased precision and reliability of the results. Surface micromachined CMUT devices with two-phase interdigital arrangement of the transducer elements were fabricated for experiments. The microchannels of different heights were modeled by attaching the adhesive tape of various thicknesses at the sides of the CMUT devices to

* Corresponding author: Darius Viržonis. Kaunas University of Technology, Panevezys institute, LT-35212 Panevezys, Lithuania. Electronic mail: darius.virzonis@ktu.lt.

† Galius Vanagas: Kaunas University of Technology, Panevezys institute, LT-35212 Panevezys, Lithuania. Rimantas Kodzius: King Abdullah University of Science and Technology (KAUST), Computer, Electrical and Mathematical Sciences and Engineering Division (CEMSE), Thuwal 23955-6900, Saudi Arabia. Envirola Project, Thuwal 23955-6900, Saudi Arabia.

form the bead for the liquid. The drop of the transformer oil was put in the microchannel before attaching the upper microchannel wall. Afterwards the electromechanical impedance of CMUT was measured with the network analyzer to evaluate the CMUT operating regime.

The finite element analysis showed that when the microchannel cross-section dimensions are comparable with the CMUT cell dimensions, the transducer is subjected to the pressure waves bouncing between the microchannel walls and comparatively compliant CMUT membranes. This situation causes the distortion of the CMUT operating regime and has to be accounted during the design and use of CMUT-coupled microfluidic devices. Corrections for the amount of the near-field distortions are to be made already during the design phase.

We found significant relationship between the CMUT membrane motion characteristics and the microchannel height. For example: CMUT having 5 MHz central frequency in immersion exhibits the broad band frequency spectra when in normal operating conditions, but the central frequency decreases by 60-70 % with decreasing the microchannel height from 400 microns to 2 microns, and the frequency band becomes narrower. It is also shown here how the simulated results can be experimentally verified.

INTRODUCTION

Microfluidic technology provides benefits such as requirement for very low sample and reagent volumes. Further advantages include fast analysis, low dead volume, and low power consumption enabling the integration of multiple processing steps (reagent storage, sample preparation, fluid handling, separation, detection and readout). The integration allows automation, limiting user interaction which results in higher reproducibility and accuracy of experimental results. [1-3]

For process automation and desirably integration into microfluidics system, fluids can be actively moved with the help of micropumps. The micropumps can be roughly divided into two categories: mechanical pumps with moving parts (piezoelectric, pneumatic, thermopneumatic, electrostatic, electromagnetic, bimetallic shape memory alloy pumps) and non-mechanical pumps without moving parts (such as electrokinetic, magnetohydrodynamic, electrochemical, acoustic-wave, surface tension and capillary micropumps). [4]

Microfluidic devices are found in a wide application range in biochemistry, genetics and other health sciences due to their usefulness for lab-on-a-chip systems and a great potential as portable point-of-care (POC) medical diagnostic devices. [2, 5-8] However, fluid micromanipulation and pumping methods are still in the developmental phase and commercially available micro pumps require better integration potential. Moreover, long and narrow pump-sensor interconnects and low fluid pumping rates result in a laminar flow with a corresponding decrease in the microchannel-fluid interaction rate. Therefore, despite several emerging microchannel-based sensing methods, most of the microchannel devices are more suitable for measurements as complex laboratory systems rather than easy-to-use high throughput POC units. There are several fluid manipulation technologies known that are conceptually suitable for direct integration into microfluidic devices. [4] One of the most appropriate methods with the potential of coupling actuation and sensing is the acoustical streaming.

Several kinds of microchannel-integrated acoustical streaming devices employing piezoelectric excitation of flexural plate waves [9, 10], as well as interdigitally arranged ultrasound transducer exciting directional Scholte waves [11-13] have been demonstrated so far. The CMUT platform is thought to be the most versatile for microchannel integrated acoustical streaming since the same device can be used for pumping, mixing and sensing a number of parameters of the fluid. [12-14] Considerable fluid transportation rates of interdigital CMUT device have been demonstrated by McLean and Degertekin. [11, 12] However, energy transfer effectiveness to the fluid stream was rather low.

Fluid Manipulation by Interfacial Acoustic Waves

Acoustical streaming of liquids near the boundary with solid [15] is one of the convenient ways of pumping and mixing microfluids. Interfacial acoustic waves, traveling at the solid and fluid interface, can be used for this purpose with the best potential of compact integration of entire microfluidic system.

There are three inter-related theories describing acoustic waves propagating at the interface of two different media. The first one, describing the wave propagating at the interface of the solid body and the vacuum has been mathematically described by Rayleigh in his fundamental work. [16] This kind of waves is referred to as surface acoustic waves (SAW) and was named after their discoverer as Rayleigh waves. As usual, SAW are characterized as waves having both longitudinal and transverse motion, traveling on the surface of elastic material, while the material particles move in elliptical orbits, and the amplitude of movement decays exponentially with depth increace. The coupling between the particles of the solid involved to the wave motion and any media other than vacuum, affect significantly the amplitude, attenuation and velocity of the wave, and in this way enable SAW sensing properties of the interfacing media.

When the propagating SAW meet another type of the interface, for example solid and fluid interface or solid and solid interface, the wave energy may be converted to other types of acoustic waves. So the case of wave propagating at the interface of two solids was described by Stoneley as an extension to classical Rayleigh's theory. [17] The theory was further extended by Scholte who added the mathematical explanation of the waves presented at solid and fluid interface. [18, 19] Scholte waves are of particular interest in microfluidics, because they can be utilized both for sensing and fluid manipulation. Interfacial waves are usually excited and captured by interdigitated (IDT) electromechanical transducers. The role of transducers here is to convert electrical signals to mechanical vibrations and vice versa. The idea of using IDT's is shown in Figure 1. Transducer elements ("fingers") are considerably flat and arranged so that they can be excited by Tx1(t) and Tx2(t); they are preferably narrow band harmonic electrical signals different in phase:

$$Tx1(t) = A\sin(\omega t)\,;\, Tx2(t) = A\sin(\omega t + \varphi),$$

where ω and φ are angular frequency and phase shift, correspondingly.

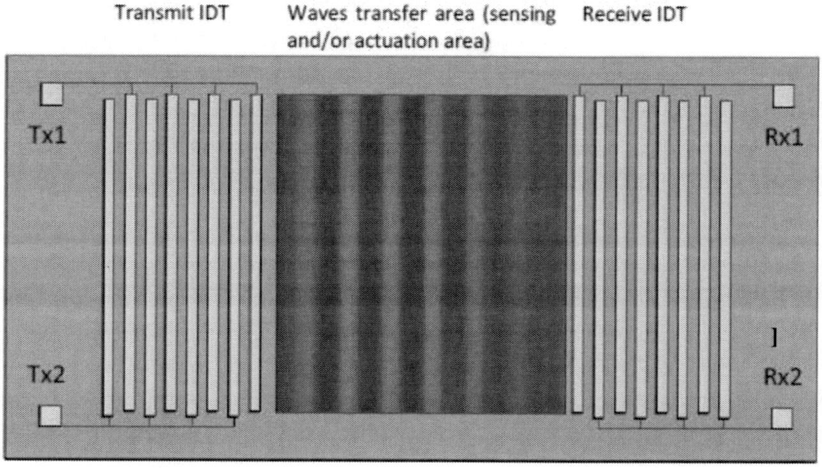

Figure 1. Using of IDT's for SAW transmit and reception.

Exposed to the Tx1 and Tx2, transducer fingers change their mechanical dimensions according to the value of the excitation signal, so they introduce mechanical waves to the substrate. The excitation of the interface waves can be unidirectional or bidirectional. In bidirectional case the waves are propagated to the left and right sides from IDT with equal energies, and unidirectional excitation allows the control over the wave propagation directness.

Figure 2 illustrates the bidirectional excitation in the $\varphi = 2\pi$ case, where the mechanical deformation of the transducers (shown in not-to-scale cross-section in Figure 2 b) corresponds to the time moment t, shown by the vertical dashed line in Figure 2 a. For effective bidirectional excitation mechanical distance between the same phase fingers has to be equal to the wavelength λ of a target wave, and fingers of other phase are to be placed exactly in the middle, i.e. at $\lambda/_2$.

The result of this arrangement is the constructive interference between the waves excited by Tx1 and Tx2 evenly on both sides of IDT, so bidirectional excitation occurs.

Unfortunately, symmetrical arrangement of IDT fingers does not provide the means for controlling the directionality of excitation: change of the φ value to either side will only decrease the excitation efficiency. To control over the directioness of the wave excitation, one needs to arrange transducer fingers asymmetrically.

One of the possible arrangements is shown in Figure 3 b: a not-to-scale cross-section of an IDT fingers. The $\lambda/_4$ spacing between Tx1 and Tx2 fingers and corresponding φ value will create the conditions for constructive interference between two waves on one side from IDT and destructive interference on the opposite side.

The side of constructive interference will change with respect of the polarity of a phase shift: right-side excitation will occur if $\varphi = \pi/_2$, and left-side excitation will occur if $\varphi = -\pi/_2$. However, only infinite number of finger pairs would allow perfect control over the wave directness. In practice, some of the wave energy will not be cancelled by the destructive interference, because the real frequency spectra of the oscillations will be broader than single frequency.

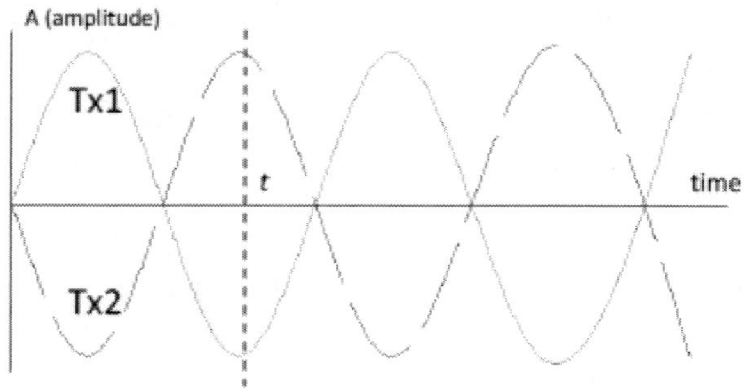

a

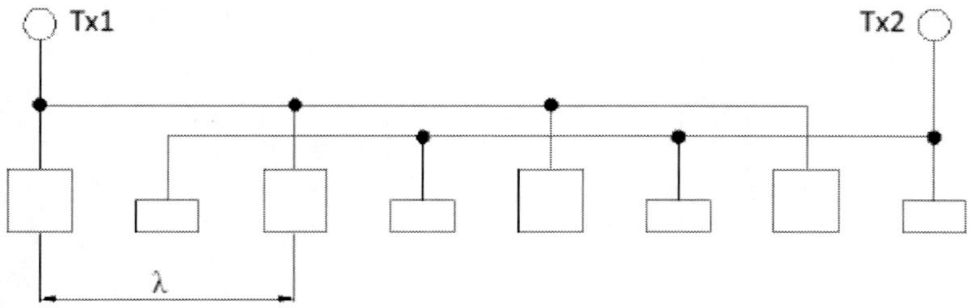

b

Figure 2. Arrangement and excitation of IDT in bidirectional case; (a) – the excitation signal; (b) – the arrangement of the interdigitated transducer elements.

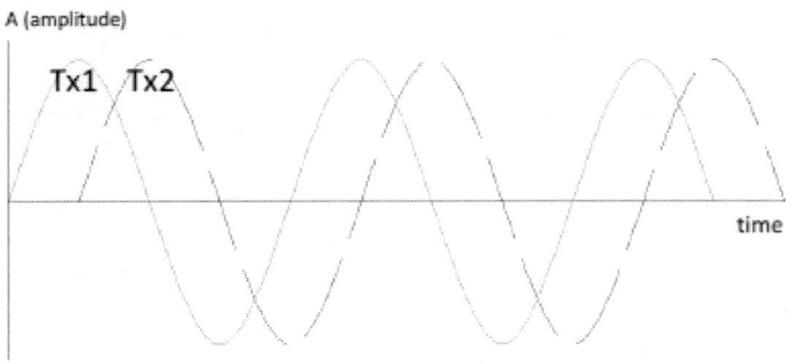

a

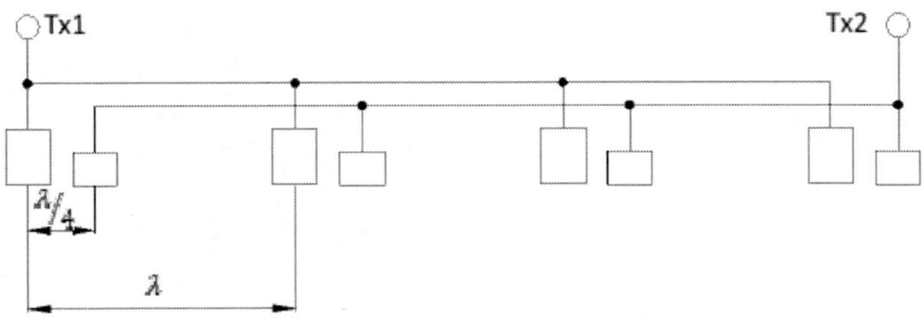

b

Figure 3. Arrangement and excitation of IDT with the directness control ability; (a) – excitation signal; (b) – the arrangement of the interdigitated transducer elements.

The electronic control of the interface wave excitation direction makes the concept very interesting for microfluidics, since traveling interface wave (particularly Scholte wave) can be utilized in reversible pumping and mixing of analyte without any additional electromechanical arrangements.

The Concept of Capacitive Micromachined Ultrasound Transducers

As described in previous section, electromechanical transducers play essential role in conversion between the mechanical and electrical energy domains: exciting and receiving interfacial acoustic waves in microchannels. Typical solution for electric-to-mechanic and vice versa conversion functionality is utilization of reversible piezoelectric effect, which is a property of some natural and artificial materials. The most important feature of the piezoelectric effect is the change of the material crystal lattice dimensions due to the applied electrical field (known as reversed piezoelectric effect) and generating of the electrical field by the mechanically deformed crystal lattice (known as direct piezoelectric effect). Widely used piezoelectric materials are single-crystal piezoelectrics (quartz, lithium niobate), piezoceramics (lead zirconate titanate) piezocomposites, which may contain any of piezoelectric materials embedded in the matrix of polymer material. Direct piezoelectric effect, discovered by Pierre and Jacques Curie in 1880, was complemented by the discovery of inverse effect by Gabriel Lippmann in 1881. The invention of ultrasonic underwater sonar by Paul Langevin and Constantin Chilowski in 1916 - 1917 sparkled the practical interest in this effect. Corresponding development of echosonography for non-destructive testing and visualization of technical and medical objects in the second half of 20th century lead to the development of advanced types of piezoelectric ultrasonic transducers, containing hundreds of electromechanical elements, enabling dynamic focusing and electronic scanning over two-dimensional and three-dimensional space.

Development of micromachining technologies in the last four decades of the 20th century provided conceptually new fabrication paradigm for microscopic sized electromechanical devices (known as Micro Electro Mechanical Systems – MEMS), and these technologies became the new object of inspiration for ultrasonic engineers. This resulted in ultrasonic transducers produced by micromachining technologies, known as Micromachined Ultrasound Transducers - MUT. The main advantage of micromachining is to break the sub-millimeter barrier of traditional machining and in this way to enable the integration of more transducer elements to less space and therefore to gain an image quality and informative content. Furthermore, micromachining enabled easy integration with electronics, because an electromechanical part is made using the technology, similar to established processes for microelectronics (like CMOS) fabrication. However, utilization of known piezoelectric materials for MUTs is complex, costly and not always desirable due to the incomplete compatibility with fabrication processes of microelectronics. Therefore the concept of eletromechanical energy conversion through electrostatic interaction was introduced in the field of ultrasonics.

The concept of electrostatic micromachined ultrasound transducers also known as capacitive micromachined ultrasound transducers – CMUT was invented 1994-1996 by Haller and Khuri-Yakub [20] and quickly resulted in boom of research activities.

In contrast to piezoelectric method of electromechanical conversion, no materials with electrically related crystal lattices are needed for CMUT: the electrostatic method of electromechanical conversion is based on the Coulomb interaction between two conductive plates, which are electrically insulated from each other, and one of them can move in respect of the other. The conceptual sketch of a single CMUT cell is illustrated in Figure 4. An elastic perimeter-clamped membrane is suspended above the substrate by electrically isolating ring-shaped post. The closed cavity is formed between the top of the substrate, bottom of the membrane and inner surface of the supporting post.

Vacuum is preferred inside of the cavity to decrease damping and improved electrical isolation. Electrical coupling with this comparatively simple mechanical structure is done by the pair of electrodes, one of which is superposed with the substrate and remains fixed during the operation, while the other (top) electrode is coupled with the membrane, and can move due to membrane deflection. When electrical field is created between the electrodes, Coulomb force attracts electrodes to each other and is offset by the membrane's stiffness. The result is conversion of supplied electrical energy to membrane's mechanical motion and mechanical energy of membrane's equivalent spring.

When the membrane is deflected by the outside pressure, the charge stored between the electrodes changes, causing the change in receiver current. This results in conversion of mechanical energy into electrical signal. Simplified electromechanical scheme of described system is presented in Figure 5: mechanical part of a CMUT cell is represented by the mass-spring-damper system, the electrical part – by capacitance, voltage and resistance.

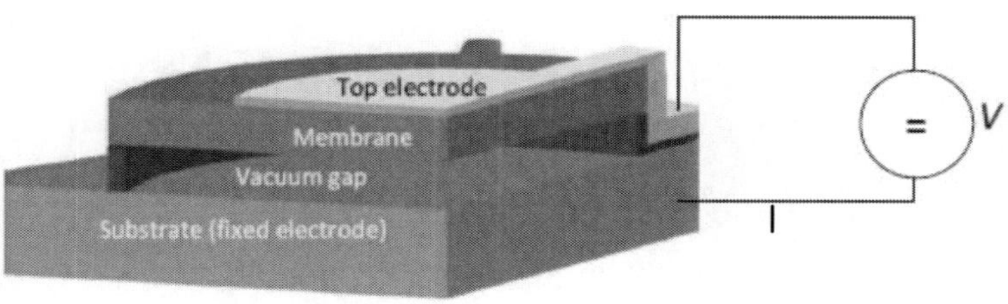

Figure 4. Conceptual sketch of quarter – sectioned CMUT cell; the voltage source V on the right creates the electrical field, which deflects the membrane.

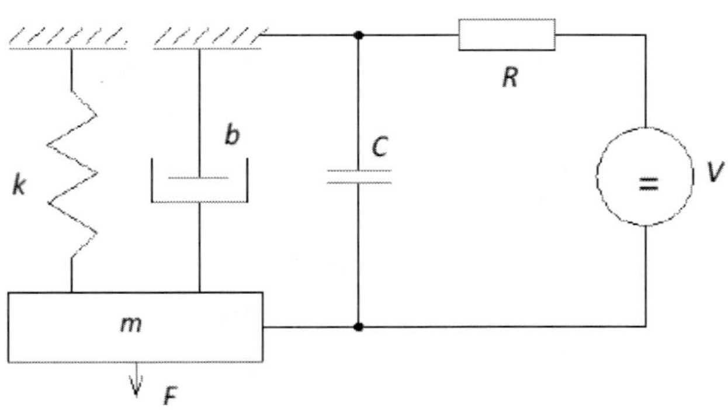

Figure 5. Simplified electromechanical schematic of CMUT; in mechanical part: *m* stands for the moveable mass of the membrane, *k* stands for membrane's equivalent spring constant, *b* stands for mechanical losses, *F* stands for the net force; in electrical part: *C* stands for the cell electrical capacitance, *R* stands for electrical losses, and *V* stands for the driving voltage.

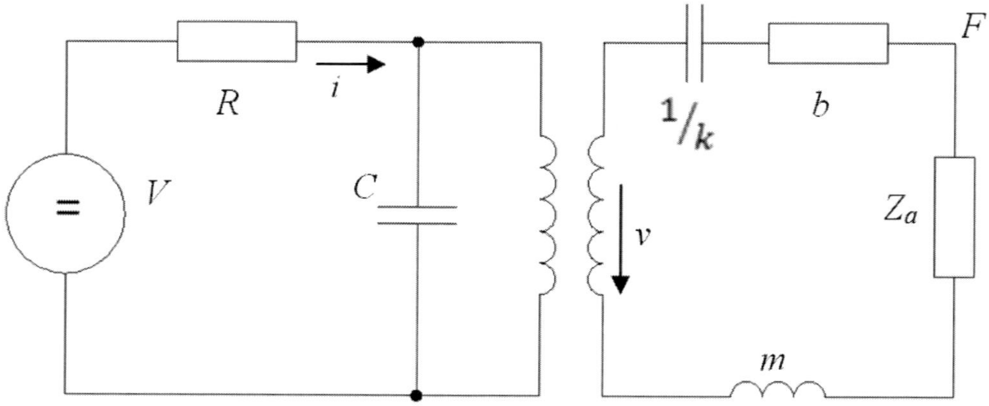

Figure 6. Simplified equivalent circuit of CMUT.

The capacitance C in Figure 5 is non-linear, as it depends on the distance between the electrodes ("electrical gap"). The gap changes when the membrane moves. Let us assume electrical gap to have uniform vacuum properties and denote it as g. This is not the case in reality, because electrical gap is only partially vacuum, while the rest part is the membrane material, according to the concept shown in the Figure 4.

The gap is essential CMUT property, which belongs exclusively neither to electrical nor mechanical domain, because it can be changed evenly by changing the electrical energy (electrical charge) or the mechanical net force acting over the membrane. One of the most appropriate methods of CMUT mathematical description is equivalent circuit method, which is widely applied by many research groups. A simplified version of CMUT equivalent circuit for conversion from voltage to force is shown in the Figure 6. The mechanic and electric energy domains in this circuit are separated by the ideal transformer, which is introduced to reflect the non-linear behavior and belonging to both energy domains of the capacitor C. The mechanical part to the right of the circuit in Figure 6 inherit notations from the mechanical schematic diagram in the left part of the Figure 5. Thus the capacitance $^1/_k$ simulates the spring behavior, the resistance b is equivalent to the mechanical losses, inductivity m stands for the mass, v stands for the velocity of the gap change, Z_a is the impedance of the auxiliary medium, and F denotes the net auxiliary force.

The balance equation for the electrical domain, according to Ohm's law:

$$\frac{(V - V_C)}{R} = i; \left(V - \frac{Qg}{\varepsilon A}\right) - i = 0.$$

Here V_C is the voltage over the capacitor C; Q is the charge stored in capacitor C; ε stands for the electric permittivity of the gap, and A is the effective area of the electrodes.

Mechanical force F_C over the capacitor's C gap, developed by the stored charge, can be found according to the Coulomb's equation:

$$F_C = \frac{Q^2}{2\varepsilon A}.$$

Balance equation for the mechanical domain can be found using Kirchhoff's law:

$$F_C + bv + m\dot{v} + k(g - g_0) - F = 0.$$

Here g_0 is the gap size when the membrane is in rest position, i.e. when both the stored electrical charge and the net auxiliary force are equal to zero.

In order to join these expressions into a system of differential equations we will introduce the following conventions: $x_1 = Q$; $x_2 = g$; $x_3 = v$. Then, assuming auxiliary impedance and net force are zero, CMUT can be described as following state-space system:

$$\left.\begin{aligned} \dot{x}_1 &= \frac{1}{R}\left(V - \frac{x_1 x_2}{\varepsilon A}\right) \\ \dot{x}_2 &= x_3 \\ \dot{x}_3 &= -\frac{1}{m}\left(\frac{x_1^2}{2\varepsilon A} + bx_3 + k(x_2 - g_0)\right) \end{aligned}\right\}.$$

Unfortunately, this system does not have explicit analytical solution. [21] By solving it numerically one can find CMUT's transient functions $Q(t)$, $g(t)$ and $v(t)$, from which some critical parameters, for instance the resonance frequency can be determined.

The Table 1 of the CMUT design parameters is given to illustrate the application of above described mathematical model.

One of the model outputs $g(t)$, obtained through described design parameters, is shown in the Figure 7.

Part a of the Figure 7 shows the transient response of the simulated gap, which is equivalent to the membrane's (assumed to be represented by the parallel plate) displacement. This response was found assuming the 40 ns duration and 50 V amplitude impulse of the voltage source V. Part b shows the power spectral density of the function $g(t)$, thus giving an estimate of the resonance frequency, which is 17.7 MHz for given design parameters.

Table 1. Design parameters for illustration of the simplified CMUT mathematical model

Parameter	Calculation	Symbol	Value
Properties of the membrane material: • Elasticity modulus • Poisson ratio • Mass density		 E ν ρ	 150 GPa 0.7 $2.33 \cdot 10^3\ kg/m^3$
Membrane radius		r	$2 \cdot 10^{-5}\ m$
Membrane thickness		h	$10^{-6}\ m$
Effective area of electrode	$A = \pi r^2$	A	$1{,}257 \cdot 10^{-9} m^2$
Vacuum dielectric constant		ε	$8{,}85 \cdot 10^{-12}$
Nominal gap (at rest)		g_0	$1{,}5 \cdot 10^{-7} m$
Moveable mass (electrode mass is disregarded)	$m = \rho A h$	m	$2{,}928 \cdot 10^{-12} kg$
Mechanical loss coefficient		b	$2 \cdot 10^{-5}$
Membrane's equivalent spring constant	$k = \frac{192\pi D}{r^2}$; $D = \frac{Eh^3}{12(1-\nu^2)}$	k	$3{,}70 \cdot 10^4$
Electrical losses		R	$1\ k\Omega$

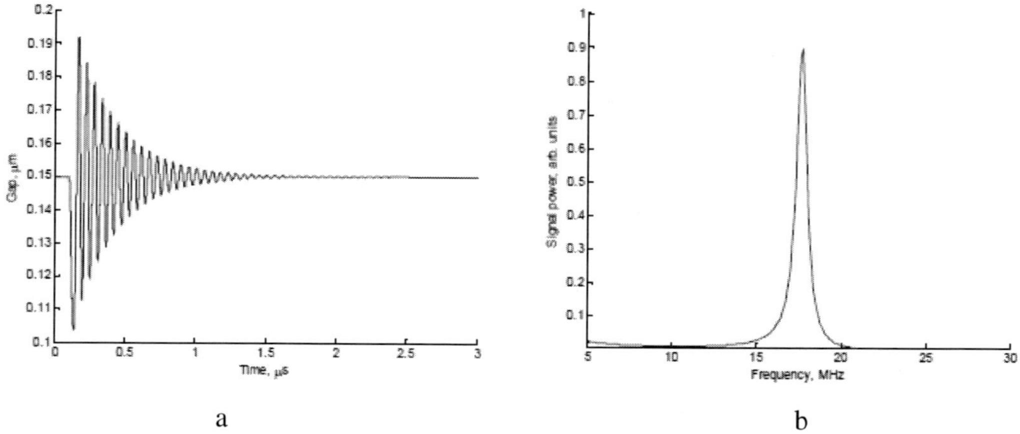

Figure 7. Calculated transient response of the CMUT cell's effective gap to the 50 V amplitude and 40 ns duration impulse (a) and power spectral density found from the transient response (b).

The approximate value of the membrane's free oscillations frequency can also be found by simple resonance frequency calculation for undamped mechanical spring-mass system:

$$f_0 = \frac{1}{2\pi}\sqrt{\frac{k}{m}}.$$

This equation gives slightly different f_0 value of 17.9 MHz for given design parameters.

If CMUT is operated in medium other than vacuum, for example in liquid, Z_a and (or) F by Figure 6 cannot be disregarded due to viscous and elastic interaction with the medium. In absence of the static auxiliary pressure, the radiation impedance can be regarded as the dominating mechanical load over the membrane. It was shown that radiation impedance of the piston transducer can be found by using following form of Rayleigh-Sommerfeld model: [22]

$$\frac{Z_a}{\rho_a c A} = 1 - \frac{[J_1(\omega r/c) - iS_1(\omega r/c)]}{\omega r/c}.$$

Here ρ_a is the density of the medium; c is the wave propagation velocity in medium; J_1 and S_1 are type 1 Bessel and Struve functions. Figure 8 shows the calculated transient response of CMUT after the radiation impedance in water ($\rho_a = 1000\,kg/m^3; c = 1500\ m/s$) is introduced to the model. The nonperiodic nature of the response is obvious; the power spectral density in damped conditions becomes the function both of the excitation pulse duration and of the membrane properties. Presented transient response was calculated for 50 V and 120 ns excitation pulse, resulting in 11.6 MHz central frequency with approximately 10 times wider fractional bandwidth than in the case of free oscillations. The bandwidth can be further increased by shortening the excitation pulse; however the amplitude of the membrane motion will decrease. The described simplified model does not reflect many of the CMUT features, for example equivalent spring constant decrease and transformer turns ratio change due to applied electrical field. More advanced mathematical models of CMUT can be found in many references. [23, 24] However, the presented model still can be used to study the collapse behavior of CMUT thus typical for electrostatic MEMS. The equilibrium between the membrane's spring force and the Coulomb force can become unstable as the spring force has linear and Coulomb force – second order dependency on the gap value. Stability analysis of Coulomb/elastic forces equilibrium is one of the methods to find the collapse voltage, [21] which is again one of most important CMUT properties, determining the range of effective operating voltage.

The net force F (acting over the membrane and directed away from the fixed electrode) if any auxiliary forces and impedance are zero, can be expressed as follows:

$$F = -\frac{\varepsilon A V^2}{2g^2} + k(g_0 - g);$$

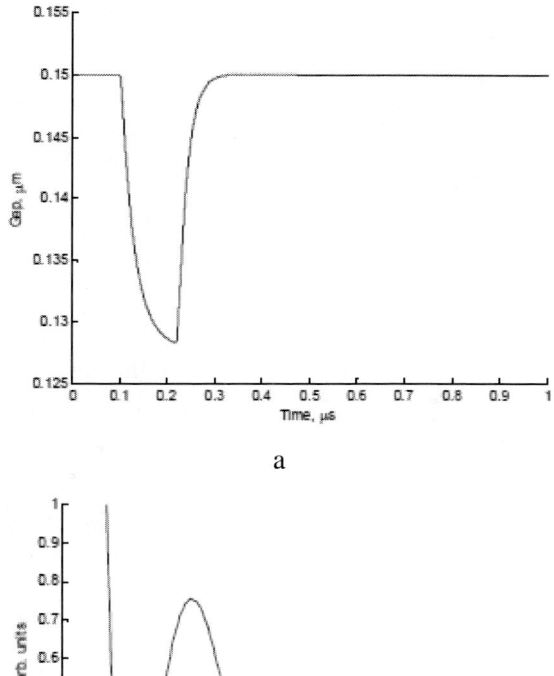

a

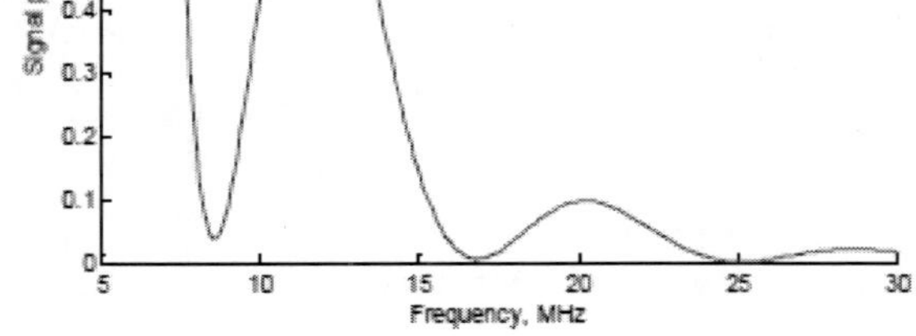

b

Figure 8. Calculated transient response of the damped CMUT membrane; (a) effective gap as a function of the time and 50 V/120 ns impulse; (b) power spectral density of the transient response.

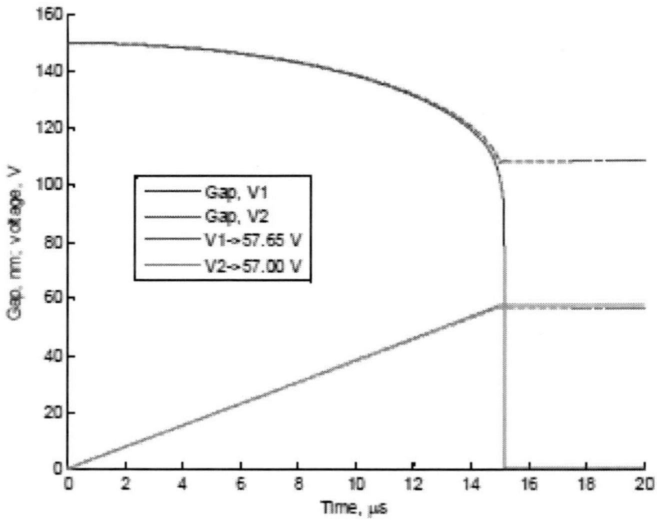

Figure 9. Calculated gap as a function of the voltage and time.

In equilibrium conditions $F = 0$. If small deviation δg is added to the gap value, corresponding change of the net force can be expressed as:

$$\delta F = \left.\frac{\partial F}{\partial g}\right|_V \delta g.$$

Here important point to note: is if a small decrease of the gap induces increase of the net force, the operating point is stable, since increase of the net force will tend to restore the gap. If δF is negative due the small decrease of the gap, the operating point is not stable, because the net force will further decrease the gap value.

From the above expressions we can derive the equation, which will lead us to the expression of the stability condition:

$$\frac{\delta F}{\delta g} = \frac{\varepsilon A V^2}{g^3} - k;$$

from where the stability condition is $\frac{\delta F}{\delta g} < 0$ or $k > \frac{\varepsilon A V^2}{g^3}$. If all the parameters describing operating point are locked except the gap and voltage, the critical voltage, causing the transition from stable to unstable operating point exists. This voltage is called the collapse voltage V_{coll} and can be found from the stability condition expression. We can estimate the critical gap, which corresponds to the collapse voltage:

$$k > \frac{\varepsilon A V_{coll}^2}{g_{coll}^3};$$

It is easy to find the critical gap value $g_{coll} = \frac{2}{3} g_0$; corresponding collapse voltage can be found for this gap value:

$$V_{coll} = \sqrt{\frac{8kg_0^3}{27\varepsilon A}}.$$

In the design parameters given in Table I, we can calculate the value of V_{coll} to be 57.65 V. Using described system of differential equations we can find the gap as a voltage and time function as shown in the Figure 9.

The voltage is increased from 0 V to reach the steady state values of $V_1 = V_{coll} = 57.65\ V$ and $V_2 = 57.00\ V$. It can be seen that CMUT collapse occurs (the gap value abruptly reaches zero) in V_1case, but not in the case of V_2, which does not reach the collapse voltage.

MATERIALS AND METHODS

Finite element analysis (FEA) and experimental evaluation of CMUT operating characteristics were used to analyze the operating parameters of CMUT in a microchannel. Two separate models of the ultrasonic wave propagation in the microchannel and CMUT membrane dynamic behavior were created using a commercially available FEA modeling package (ANSYS, Inc., Canonsburg, PA). Those were simulations of 1) CMUT induced harmonic acoustic waves in the microchannel and 2) transient response of CMUT cell under the near-field operation conditions.

CMUT Fabrication and Testing

For experimental purposes surface micromachined CMUTs with silicon nitride ($E=1.3\cdot10^{11}$ Pa; ν=0.7) membranes having 38 μm effective diameter and 0.8 μm membrane thickness were designed for 10 MHz resonance in air and 5 MHz central frequency in immersion. Highly doped silicon <100> substrate with a 0.5 μm thermal oxide layer was used for CMUT micromachining. Schematic representation of the technological process sequence is shown in the Figure 10. Cavities of CMUT cells were etched down to the substrate using photo resist as a mask with two step etching process: anisotropic reactive ion etching followed by the anisotropic buffered oxide etching (Figure 10 b). The sacrificial chromium film was deposited on top of the resist mask (which was not removed after the cavity definition) and later used to pattern the sacrificial film using the lift-off technique (Figure 10 c). Three thicknesses of the sacrificial film were used to produce devices with different vacuum gap height: 150 nm for type A, 80 nm for type B and 100 nm for type C.

The first 400 nm layer of structural silicon nitride film was deposited via plasma enhanced vapor deposition using silane and ammonia mixture (Figure 10 d). [25] During the next fabrication step, the sacrificial etching holes were formed in the structural silicon nitride film via isotropic reactive ion etching followed by the sacrificial film removal using a wet anisotropic etching (Figure 10 e). After drying, devices were sealed by depositing an additional 300 nm silicon nitride film (Figure 10 f).

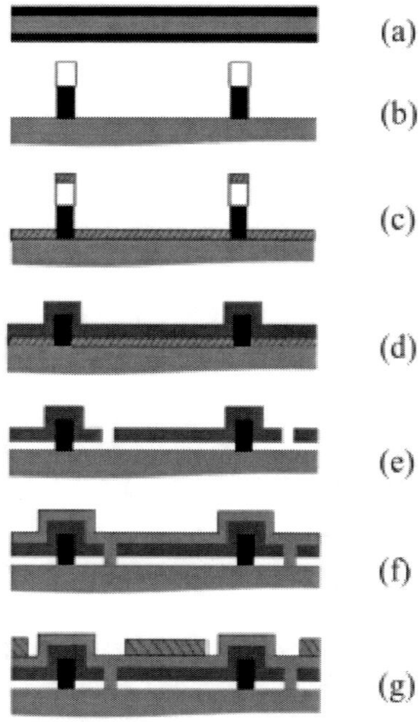

Figure 10. Pictorial representation of the technological process used in CMUT fabrication: (a) – silicon wafer covered by 0.5 μm thermal oxide layer; (b) – cavity etching down to the substrate; (c) – sacrificial chromium film deposition; (d) – first structural nitride layer deposition; (e) – sacrificial etching vias opening and sacrificial etching; (f) – second structural nitride layer deposition and sealing of cavities; (g) – electrode deposition and patterning.

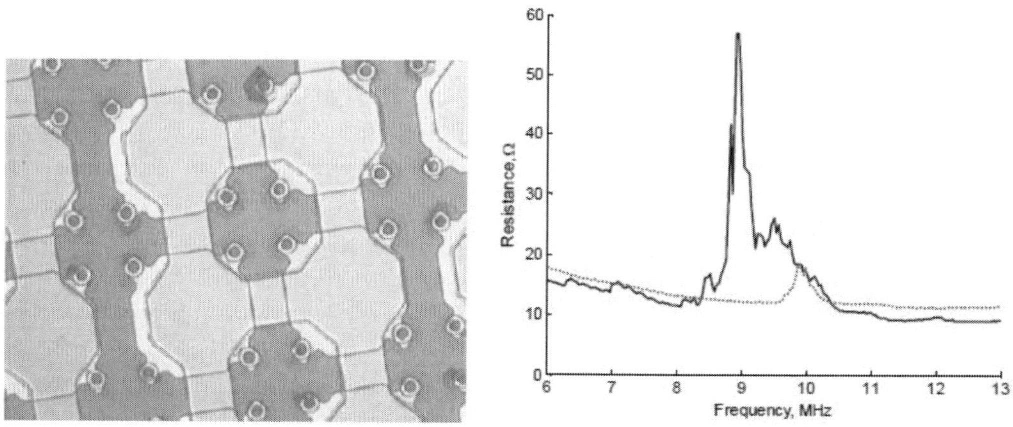

a b

Figure 11. a) – fabricated CMUT cells; b) – air resonance tests at 0 V (dotted line) and 55 V (continuous line).

Finally, 200 nm thick top electrodes and bonding pads for bottom electrodes were machined by titanium-gold evaporation and lift-off processes (Figure 10 g). Three types of CMUT devices, all having the same octagonally shaped membranes of 38 μm, differing in the sacrificial film thickness and the corresponding vacuum gap height, were fabricated. Type A had the vacuum gap of 150 nm while type B - 80 nm and type C - 100 nm. Optical micrograph of the fabricated type A CMUT structure is shown in Figure 11 a, and the measured resonances in air are shown in Figure 11 b.

The CMUT testing was done by connecting it to the Agilent 4395A network analyzer with the impedance measurement kit in the resistance measurement mode and the auxiliary DC source of bias voltage.

The bias voltage ranging from 0 to 120 V was applied during the tests, giving the control over the neutral deflection of the membranes and therefore changing the amplitude of CMUT membranes oscillations and the resonance frequency. The resistance spectra measured at 55 V bias (shown at Figure 11 b as continuous line) is exhibiting the air resonance approximately at 9.1 MHz, while the resonance measured in 0 V bias voltage case gave the 9.8 MHz value, which is quite close to the design target.

Microchannel Fabrication

Two ways of fabricating the microchanels were used. Microchannels of 50 – 3000 μm height were modeled by attaching strips of the adhesive polyethylene films of different thickness to the sides of the CMUT device. After this modification, SEM images were acquired to measure the actual channel wall height.

Alternatively, for smaller channel height values, ranging from 2 to 50 μm, PDMS molding technique for the microchannel fabrication was used. Figure 12 a shows the SEM image of CMUT device with the adhesive film walls mounted, and Figure 12 b shows the die with two CMUT devices mounted with 50 μm height PDMS channel.

After the SEM wall height measurements, devices were taken to the resonant frequency and electromechanical impedance measurement site. The complete microchannel, consisting of CMUT die at the bottom and the adhesive tape walls, was filled with the transformer oil to ensure fluid environment; the device then was capped with a piece of PVC sheet. This arrangement resulted in a wide variety of CMUT operating conditions thus imitating the heights of the microchannel from 50 μm to 3 mm. Two A type, one B type and two C type devices were involved in the experiments.

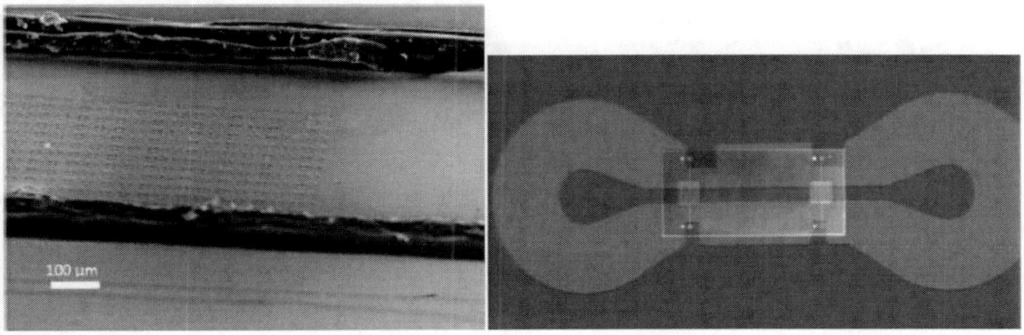

Figure 12. Left: SEM image of CMUT structure as well as the attached film to mimic the microchannel wall. Measured height of the microchannel vertical wall shown in the image was 89 μm. Right: optical image of the CMUT die with two interdigital devices mounted with the PDMS microchannel device.

Simulation of CMUT Induced Interfacial Wave in the Microchannel

This model was used to determine the relationship between the operating frequency of CMUT and the energy transfer to the interfacial wave, intended to induce the fluid manipulation in the sensor area. Interdigital arrangement of two-phase CMUT elements with $\lambda/4$ finger spacing between the phases and λ-spaced finger groups (see Figure 3) was used as a starting point.

Figure 13 shows a pictorial representation of a two-dimensional model for the investigation of the acoustic wave propagation in the microchannel. PLANE 42 elements were used to divide the top and bottom structure of the microchannel and CMUT membranes. Fluid in the microchannel was divided using FLUID29 elements with FLUID129 elements at the microchannel ends to represent the infinite length.

CMUT membranes were excited using 10^{-6} N amplitude harmonic force oscillations with a 90° phase angle difference. The 1500 m/s value of acoustic wave propagation velocity was used.

Simulation of CMUT Cell Transient Processes in Microchannel

In order to investigate the shift of CMUT operating point due to the reverberations of bulk acoustic waves we used the transient response FEA, which is schematically shown in Figure 14.

The top and the bottom walls of the microchannel were divided using PLANE42 elements, and the fluid-structure interface was supplemented by the interfacial FLUID29 elements.

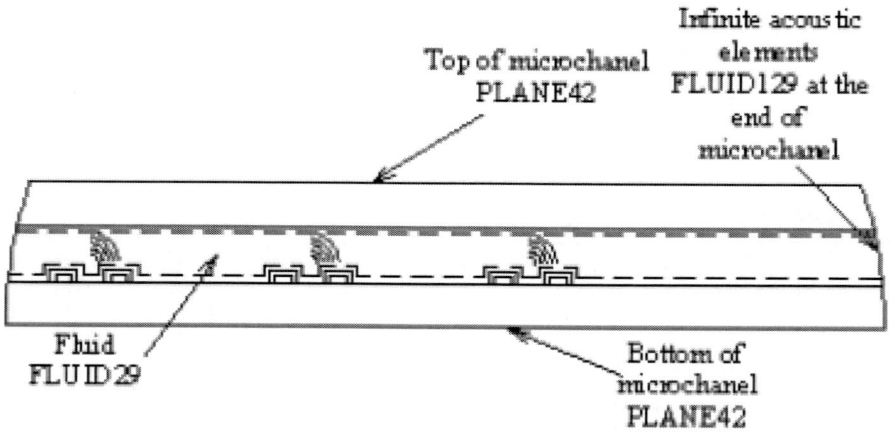

Figure 13. Pictorial representation of the finite element analysis model used to do the harmonic analysis and excitation of interfacial waves in the microchannel.

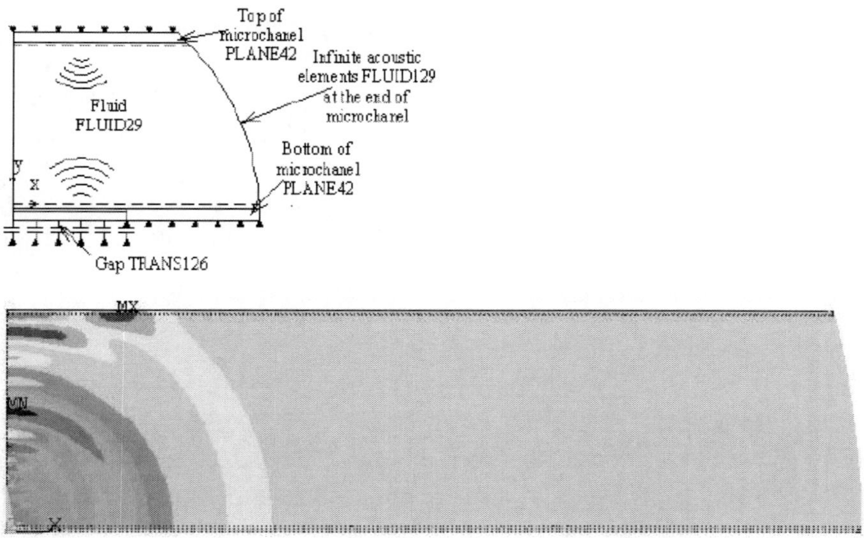

Figure 14. At the top: the structure of the finite element analysis model used to do the transient response analysis; bottom: simulated pressure wave in 100 μm height microchannel, approx at 80 ns after the membrane excitation.

Table 2. Main parameters of the model

Parameter	Notation	Value
Central frequency in immersion		7.1, 5.3, 4.0 MHz
Cell radius	a_m	15, 17.5, 20 μm
Membrane thickness	d_m	1.0 μm
Vacuum gap	d_g	0.15 μm
Isolating film thickness	d_i	0.35 μm
Electrode radius	a_e	$a_e=a_m$
Electrode thickness	d_e	0.4 μm
Isolating film material	SiO_2	$\varepsilon_i'=3.7$
Electrode material	Al	E=68 GPa, ρ=2699 kg/m^3, ν=0.35
Membrane material	Si	E=148 GPa, ρ=2330 kg/m^3, ν=0.17

The bulk of the fluid was divided using the conventional FLUID29 elements with the infinite acoustic elements FLUID129 at the side of the channel model to suppress the non-representative reflections. Here CMUT cell is represented by the membrane, the electrode structure and the vacuum gap. The membrane was separated from the electrode using the PLANE42 elements and the vacuum gap was simulated using the TRANS126 elements. The pitch of the finite elements in the network was 0.2 μm and the sampling period was 0.2 ns.

A total transient response time of 1 μs was used in simulations. CMUT membrane was excited by applying a 50 V and 40 ns impulse across the gap and the output of the model was read as the position of the central point of the CMUT cell membrane/electrode determined by the applied electrical field and/or pressure pulse echoed from the surfaces of the microchannel.

Experimental Evaluation of CMUT Operating Point

A conventional procedure of the central frequency and the electromechanical impedance measurements was used to estimate the operating point of CMUT. [27, 28] Agilent 4395A network analyzer with the impedance measurement kit was used. The microchannel height was measured by using a surface profiler and scanning electron microscope (SEM) before the experimental evaluation of the operating point and before attaching the microchannel to the transducer's die.

RESULTS AND DISCUSSION

Figure 15 shows the relationship between the simulated CMUT operating frequency shift and the corresponding excited interfacial wave attenuation. It can be seen that a significant energy loss (up to -10 dB) via attenuation of the excited interfacial wave takes place while the actual operating frequency shifts from the design point as much as 10%. This shows the simulated system being quite prone alter its operating frequency.

Figure 16 shows the output of the transient response model for the 40 μm membrane in both time and frequency domains. The 400 μm microchannel height case represents the operation of CMUT where transmitted and received pulses are separated by more than 0.5 μs thus still providing enough time for transmit and receive pressure waves to run independently. The central frequency of CMUT membrane oscillation was observed at 4.0 MHz in the frequency domain, thus representing the normal operating regime (Figure 7 b).

In the case of 20 μm microchannel height the time difference between the transmitted and received pulses is approximately 10 times smaller than the single period of the membrane oscillation in the normal regime at 26.7 ns and 250 ns, respectively.

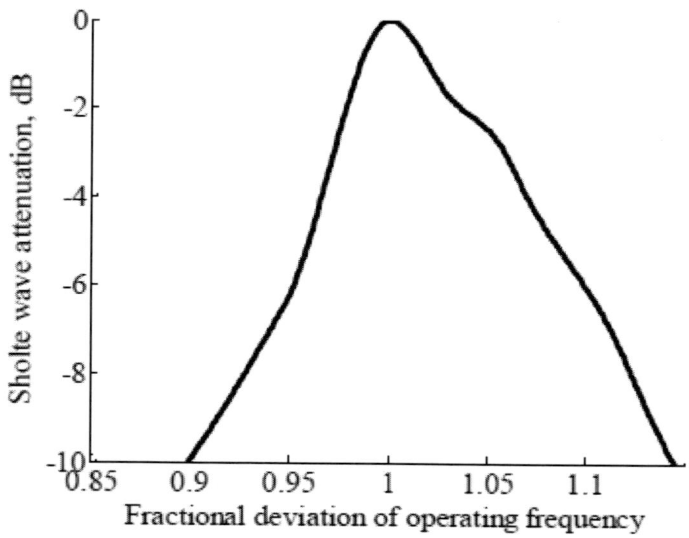

Figure 15. Attenuation of the directional interfacial wave as a function of CMUT operating frequency deviation from the design value. Here design value is unity.

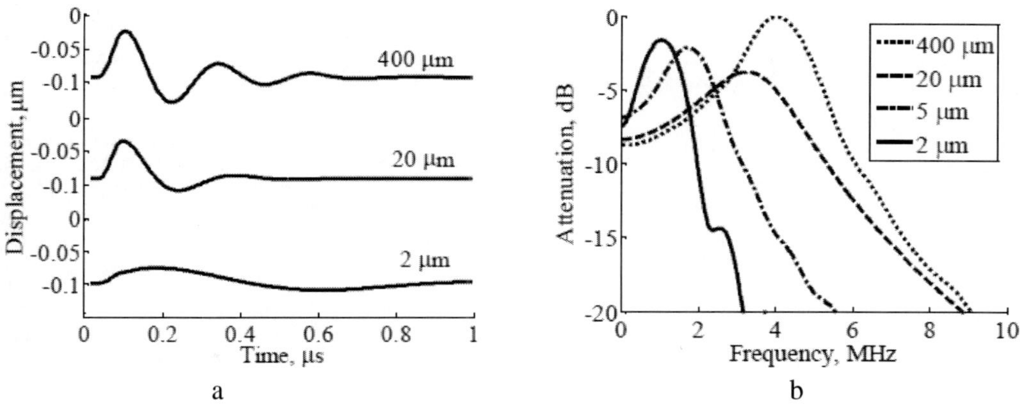

Figure 16. Displacement of CMUT membrane central point as an output of transient response analysis: (a) – time domain representation; (b) – frequency domain representation.

Here, the interaction between the membrane, which is excited by the 40 ns duration electrical pulse, and the first pressure wave reflections from the upper microchannel wall occurs, distorting the transmission. The central frequency of the membrane oscillation is less than 3.5 MHz meaning that effective operating frequency is shifted by more than 12.5 %.

In this case, frequency spectra of the oscillation can be characterized as having wider bandwidth than that in the normal regime. The bandwidth widening is expected for increased damping of the free-running membrane.

Finally, the 5 µm and 2 µm channel height cases illustrate the result of the convolution of the pulse being transmitted with many pressure reflections from both simulated surfaces resulting in CMUT membrane motion having well expressed central frequency at 1.8 MHz and 1.0 MHz for 5 µm and 2 µm channel heights, respectively, i.e. much lower than those in the normal condition operation simulations.

The relationship between the shift of the central CMUT membrane oscillation frequency and the microchannel height is shown in Figure 17 a. Three different size membranes and their corresponding central frequencies of oscillation in immersion (40 µm for 4.0 MHz, 35 µm for 5.3 MHz and 30 µm for 7.1 MHz) were simulated. The rest of the model parameters were kept constant for all the three types of membranes as shown in Table I. It can be seen that smaller membranes oscillating at higher frequencies deviate in oscillation frequency more than the larger membranes.

However, fractional deviation has an opposite behavior: the 30 µm membrane oscillation frequency decreases up to 65 % (from 7.1 MHz in normal operation to 2.5 MHz when the upper wall is within 2 µm) and the 40 µm membrane - by 74 % (from 4.0 MHz in normal operation to 1.0 MHz in 2 µm case). The 35 µm membrane frequency decrease was 70 % (from 5.3 MHz in normal operation to 1.6 MHz in 2 µm case).

This can be explained by the higher compliance of larger membranes and higher stiffness of the smaller ones. The described relationship is governed by the multiple constructive and destructive mechanical interferences between the transducer membrane and the pressure waves. The general outcome of these complex interactions can be described as the channel height dependent mechanical losses, which follows a logarithmic approximation.

Figure 17 b shows the relationship between the microchannel height and the amplitude of the oscillation of the central point of the membranes.

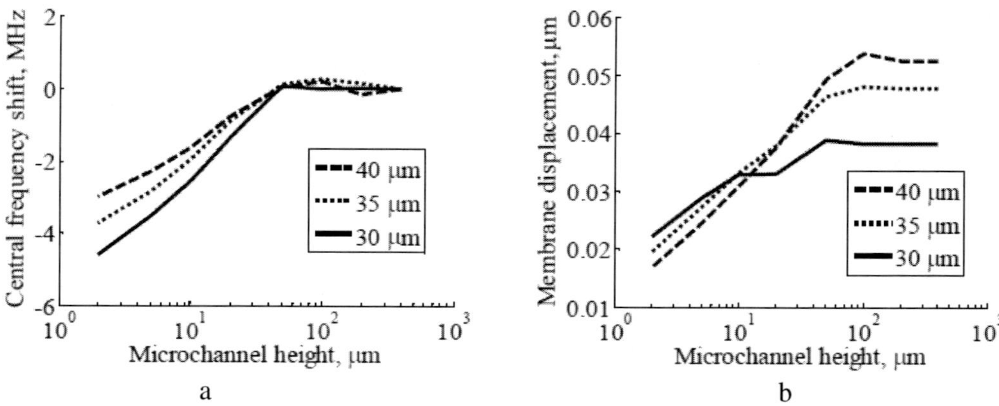

Figure 17. Shifts of the mechanical operating point of different size CMUT cells as a function of the channel height; (a) – shift of the central frequency of the membrane oscillations; (b) – displacement amplitude of the membrane central operating point.

The inflection of the relationship at ~50 μm observed in Figure 17, approximately the same for all the three membrane sizes, corresponds to the operating point where near-field effect on the oscillation frequency and amplitude becomes monotonic with respect to the decreasing microchannel height.

Information shown in Figure 17 b also confirms that stiffer membranes (with the smaller diameter) loose less of the motional amplitude due the near-field interaction; the decrease of the 30 μm diameter membrane oscillation amplitude was 42 % across the analysis interval, while a corresponding decrease of 40 μm diameter membrane oscillation amplitude was already 68 %.

Experimental measurements of CMUT operating point were performed to validate the results obtained during FEA simulations. Measured central frequencies and peak resistance amplitudes were normalized with respect to the nominal central frequency or peak resistance of the immersed device in normal operation and are shown in Figure 18. It can be seen that measured central frequency deviates from nominal when entering damped operating conditions within +5% to - 25% interval.

Correspondingly, Figure 18 a shows that the peak value of the real part of the electromechanical impedance (the resistance, which is the indirect measure of the membrane oscillation amplitude, according to [27, 28]) deviates by +60 % to – 90% from the nominal values. Thus, there is a qualitative agreement between the FEA simulated and experimental central frequency and oscillation amplitude shifts. When microchannel height decreases, so does the central frequency and peak resistance value.

We attribute the lack of exact quantitative match between the microchannel height (horizontal axis) in the simulation and measurement results to the low controllability of the measurement conditions. For example, the force used to hold the microchannel cap in place was manually controlled and is not known; we assume that the variations in holding force influenced the actual height of the microchannel due the uncontrolled deformations of the adhesive film walls and the microchannel cap itself.

Another possible reason for quantitative mismatch is the model used for FEA, which presumed the absence of the side walls of the microchannels, as they were present during the experiments and presumably introduced additional pressure reflections.

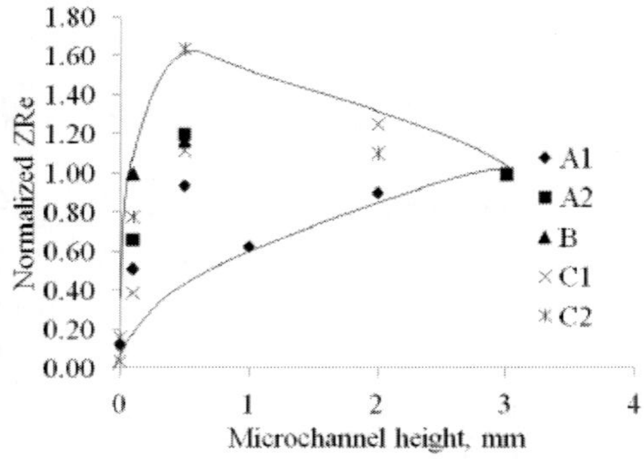

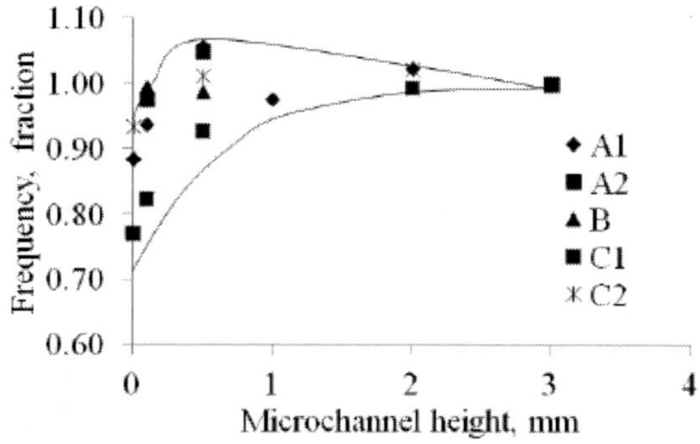

Figure 18. Experimental evaluation of CMUT operating point shift; (a) – amplitude of the membrane movement evaluated from the measured electromechanical impedance real part; (b) – deviation of the central operating frequency. The lines are added for the eye guidance only.

CONCLUSION

A qualitative agreement between the transient FEA results and the experimental measurements confirms the hypothesis about the nature of CMUT operating point shift under different microchannel height conditions. It can be proposed that dynamical load produced by the near-field reverberations has a significant impact on CMUT membrane oscillation frequency, as well as its amplitude. We conclude that in the microchannel CMUT design the frequency shift and corresponding attenuation of the effective energy transferred to the CMUT cell have to be accounted for in the design phase. Furthermore, we found significant sensitivity of the Scholte wave excitation extent towards the excitation frequency deviation, which can be roughly estimated at 0.8 dB per percent (-10 dB attenuation of CMUT-excited interfacial Scholte waves for 12 % excitation frequency deviation). Since the planar dimensions and spacing of CMUT cells and elements are determined during the design phase and cannot be changed after its fabrication, it is very important to account for the expected effective operating frequency and oscillation amplitude shifts in order to preserve the efficiency of the ultrasonic energy transfer to the microfluidic device.

ACKNOWLEDGMENT

This publication is based on work supported by Seed Fund Award "Envirola" made by King Abdullah University of Science and Technology (KAUST). A part of this research was funded by a grant (No. MIP-059/2012) from the Research Council of Lithuania.

REFERENCES

[1] X. Gong, X. Yi, K. Xiao, S. Li, R. Kodzius, J. Qin and W. Wen, *Lab on a Chip* 10 (19), 2622-2627 (2010).

[2] J. Wu, R. Kodzius, K. Xiao, J. Qin and W. Wen, *Biomedical microdevices* 14 (1), 179-186 (2012).

[3] X. Yi, R. Kodzius, X. Gong, K. Xiao and W. Wen, *Biomicrofluidics* 4 (3), 036503-036508 (2010).

[4] C. Zhang, D. Xing and Y. Li, *Biotechnology Advances* 25 (5), 483-514 (2007).

[5] Q. Cao, M. Mahalanabis, J. Chang, B. Carey, C. Hsieh, A. Stanley, C. A. Odell, P. Mitchell, J. Feldman, N. R. Pollock and C. M. Klapperich, *PLoS ONE* 7 (3), e33176 (2012).

[6] D. Eicher and C. A. Merten, *Expert Review of Molecular Diagnostics* 11 (5), 505-519 (2011).

[7] W. G. Lee, Y.-G. Kim, B. G. Chung, U. Demirci and A. Khademhosseini, *Advanced Drug Delivery Reviews* 62 (4–5), 449-457 (2010).

[8] S. Shoji and M. Esashi, *Journal of Micromechanics and Microengineering* 4 (4), 157 (1994).

[9] C. E. Bradley, J. M. Bustillo and R. M. White, presented at the Ultrasonics Symposium, 1995. Proceedings., *1995 IEEE*, 1995 (unpublished).

[10] T. Min-Chien and L. Tzong-Shyng, presented at the Nano/Micro Engineered and Molecular Systems, 2007. *NEMS '07. 2nd IEEE International Conference on,* 2007 (unpublished).

[11] J. McLean and F. L. Degertekin, presented at the Ultrasonics, *2003 IEEE Symposium on,* 2003 (unpublished).

[12] J. McLean and F. L. Degertekin, Ultrasonics, Ferroelectrics and Frequency Control, *IEEE Transactions on* 51 (6), 756-764 (2004).

[13] K. K. Park, H. Lee, M. Kupnik, Ö. Oralkan, J.-P. Ramseyer, H. P. Lang, M. Hegner, C. Gerber and B. T. Khuri-Yakub, *Sensors and Actuators B: Chemical* 160 (1), 1120-1127 (2011).

[14] A. Ramanaviciene, D. Virzonis, G. Vanagas and A. Ramanavicius, *Analyst* 135 (7), 1531-1534 (2010).

[15] W. L. Nyborg, *The Journal of the Acoustical Society of America* 30 (4), 329-339 (1958).

[16] L. Rayleigh, *Proceedings of the London Mathematical Society s1*-17 (1), 4-11 (1885).

[17] R. Stoneley, *Proceedings of the Royal Society of London.* Series A, Containing Papers of a Mathematical and Physical Character 106 (738), 416-428 (1924).

[18] J. G. Scholte, *Proceedings of the Koninklijke Nederlandse Akademie van Wetenschappen* 45 (2), 159-164 (1942).

[19] J. G. Scholte, *Proceedings of the Koninklijke Nederlandse Akademie van Wetenschappen* 45 (1), 20-25 (1942).

[20] M. I. Haller and B. T. Khuri-Yakub, Ultrasonics, Ferroelectrics and Frequency Control, *IEEE Transactions on* 43 (1), 1-6 (1996).

[21] S. D. Senturia, (Springer US, 2002), pp. 125-147.

[22] M. Greenspan, *The Journal of the Acoustical Society of America* 65 (3), 608-621 (1979).
[23] C. Meynier, D. Certon, F. Teston, N. Felix and G. Ferin, presented at the Ultrasonics Symposium, 2007. *IEEE, 2007* (unpublished).
[24] H. Kagan Oguz, S. Olcum, M. N. Senlik, A. Atalar and H. Koymen, presented at the Ultrasonics Symposium (IUS), *2009 IEEE International*, 2009 (unpublished).
[25] M. Mikolajūnas, R. Kaliasas, M. Andrulevičius, V. Grigaliūnas, J. Baltrušaitis and D. Viržonis, *Thin Solid Films* 516 (23), 8788-8792 (2008).
[26] G. Vanagas, M. Mikolajūnas, R. Kaliasas and D. Viržonis, *presented at the 10th International Workshop on Micromachined Ultrasound Transducers*, Salerno, Italy, 2011 (unpublished).
[27] F. Teston, C. Meynier, E. Jeanne, N. Felix and D. Certon, presented at the Ultrasonics Symposium, 2006. *IEEE, 2006* (unpublished).
[28] G. Vanagas, D. Viržonis and M. Mikolajūnas, presented at the Intelligent Technologies in Logistics and Mechatronics Systems, *INTELMS*, Panevėžys, Lithuania., 2008 (unpublished).

Reviewed by Prof. Yi-Kuen Lee (Mechanical Engineering and Biomedical Engineering, Hong Kong University of Science and Technology, Hong Kong) and by Prof. Valentinas Snitka (Research Center for Microsystems and Nanotechnology, Kaunas University of Technology, Kaunas, Lithuania)

In: Microfluidics
Editors: Stefano Panzarella and Walter Maroni
ISBN: 978-1-62417-345-5

Chapter 6

SELECTION OF EASILY ACCESSIBLE PCR- AND BIO-COMPATIBLE MATERIALS FOR MICROFLUIDIC CHIPS

***Kang Xiao*[1], *Jinbo Wu*[2] *and Rimantas Kodzius*[2,3,4,*]**
[1]Department of Biology, The Hong Kong University of Science and Technology, Clear Water Bay, Kowloon, Hong Kong
[2]Department of Physics, The Hong Kong University of Science and Technology, Clear Water Bay, Kowloon, Hong Kong
[3]King Abdullah University of Science and Technology (KAUST), Computer, Electrical and Mathematical Sciences and Engineering Division (CEMSE), Thuwal, Saudi Arabia
[4]Envirola Project, Thuwal, Saudi Arabia

ABSTRACT

Conventional fabrication of microfluidic chip is a complicated and time, effort and material consuming process. Consequently, due to high expenses, it has poor applicability for performing mass biological analysis by microfluidics. In this study, we report several measures to make simple, low-cost and rapidly fabricated two-dimensional (2D) and three-dimensional (3D) microfluidic chips for biological analysis. We employed various easily accessible and polymerase chain reaction (PCR) compatibility verified materials in the process of design, patterning and bonding. Firstly, we established a simple method for evaluating the PCR compatibility of various common materials, as DNA and proteins are common components or targets in biological analysis. These materials employed in fabricating microfluidic chips, include silicon, several kinds of silicon oxide, glasses, plastics, wax, and adhesives. Two-temperature PCR was performed with these materials to determine their PCR-inhibitory effect. In most cases, the addition of bovine serum albumin (BSA) effectively improved the reaction yield. We also studied the individual PCR components from the standpoint of adsorption. Most of the materials did not inhibit the DNA, although they noticeably interacted with the polymerase. Our results provide an

* Author to whom correspondence should be addressed. Electronic mail: rimantas.kodzius@kaust.edu.sa and kodzius@gmail.com.

overview of materials that are PCR-friendly for fabricating microfluidic devices. The PCR reaction, without any additives, performed best with pyrex glass, and it performed worst with PMMA. Then, based on the results of the PCR-compatible screening test, we concentrated on fabricating microfluidic chips using easily accessible materials, such as paper and wax. We successfully shortened the whole fabrication process for complicated 2D and 3D microfluidic chips from the usual several days into several minutes by using easily accessible materials and methods. We demonstrated that these types of chips have good bio-compatibility enabling their applications in many biological analyses such as PCR and DNA capillary electrophoresis. We also dramatically simplified the process of fabricating complicated 3D chips by using cyanoacrylate-based resin or adhesive wax as the bonding material to integrate chip materials such as paper, glass slides or other polymer films. With this process, a 3D microfluidic chip is achievable by vacuating and venting the chip. And the bio-compatibility and applicability of the paper-based 3D microfluidic chip was verified in the applications such as PCR, *HeLa* cell electroporation and the chemotaxis of *E. coli*. In all, our PCR-compatibility test provides good hints as to the selection of materials for microfluidic chips employed in biological analysis. Base on this, we succeeded in fabricating complicated microfluidic chips using easily accessible materials that may make it much more applicable than conventional methods.

INTRODUCTION

The fluids in the micro scale behave distinctly from those in the macro scale because many insignificant factors in the macro scale, such as surface tension, energy dissipation and fluidic resistance, become to be dominative in the micro scale system. In particular, microfluidics ensures high specificity of chemical and physical parameters, including concentration, pH, and temperature, which significantly improve reaction conditions and subsequently achieve high productivity in multiple step reactions. [1] In contrast with traditional mass volume approaches, microfluidics enables lower sample and energy consumption, allowing faster reactions for rapid detection. Therefore, microfluidics chips are widely employed in multidisciplinary research fields, including physics, chemistry, and biotechnology for the precise control and manipulation of fluids at sub-millimeter scale in lab-on-a-chip technology. [2]

Despite the high diversity of microfluidic applications, because of the limitations of realized structures and features, the selection of materials turns out to be the major barrier to implement microfluidic chips into mainstream technology for research and detection. Initial microfluidic chips were fabricated in silicon based material which was directly derived from semiconductor fabrication techniques. As the demands for various biological and chemical applications keep increasing, many factors, such as optical characteristics for imaging techniques and bio-compatibility for biologic reactions, have been taken into consideration in chip fabrication. Therefore, many novel materials and processes have been utilized. Currently, various types of materials are used in fabricating microfluidic chips, such as metals, semiconductors, glasses, plastics and others. [3]

Depending on the materials and desired structures for various applications, microfluidic chips can be fabricated in particular processes, such as photolithography; hot embossing, dry or wet etching, breaking and dicing, and thermal or anodic bonding. The most common step to fabricating microfluidic chips is photolithography, which removes designated parts of a thin film or the bulk of a substrate by using particular light irradiation, depending on the light-sensitive chemical photoresist applied, and subsequently converts a geometric pattern designed from the mask to the photoresist on the substrate. Then a series of chemical treatments are followed in order to either engrave the exposed pattern or deposit new materials in the desired pattern. A single iteration of photolithography contains several steps, including cleaning, preparation, photoresist application, exposure and developing, etching and photoresist removal, which is a time and effort consuming process, as well as requires a well-equipped clean room. [4] Therefore, the whole fabrication of microfluidic chips is a complex and difficult process. Also, the requirements for solvent or thermal processing on materials during fabrication as well as the nature of selected materials may contradict the function of biologically active components, so the bio-compatibility of the materials utilized for bio-microfluidic devices should be improved and validated to ensure the success of bio-reactions.

In this publication, we demonstrated a simple method of evaluating the PCR-compatibility of various common materials utilized in microfluidic chips, including silicon, several kinds of silicon oxide, glasses, plastics, wax and adhesives, because adsorption or inhibition of the activity of biomolecules (such as DNA and proteins) is one of the major causes for the failure of bio-reactions. Then we report several measures to make simple, low-cost and rapidly fabricated two-dimensional (2D) and three-dimensional (3D) microfluidic chips for biological analysis by bio-compatibility verified materials.

EVALUATION OF VARIOUS CHIP MATERIALS ON PCR COMPATIBILITY

Polymerase chain reaction (PCR), invented by Dr. Kary B. Mullis in 1983, is an *in vitro* thermal cycling process for amplifying DNA from template samples, [5] which earned the inventor the Nobel Prize for chemistry in 1993 and now is a powerful approach widely used in a variety of biological research, diagnosis of diseases and crime investigations. For over a decade, PCR has been studied extensively in microfluidic chips due to of the demand for minimized sample and reagent amount, reducing the overall reaction cost as well as achieving the high efficiency in reaction time and productivity of PCR realized by different microdevices. The great achievement has been made in the mature processing of microdevices (such as selection of substrate materials, surface modification, structure of reaction vessel and manipulation of sample fluid), thermocycling control, detection of amplified nucleic acid products and integration with downstream functional microdevices. [6-9] In recent years, microfluidic PCR devices have been brought into real world applications, e.g., point-of-care diagnostic devices.

Because of intensive studies and extensive applications of PCR in microfluidics, in this publication we evaluated the inhibitory affect of different microfluidic materials on PCR, which is one of the most frequently used enzymatic reactions in microfluidics. The PCR components include the DNA template, primers, DNA polymerase, dNTPs, a buffer, divalent ions ($MgCl_2$), and KCl. Due to ionic and hydrogen bonding effects, polar molecules such as DNA and the polymerase can be dramatically adsorbed by certain chip materials, which results in the failure of PCR in micro scale studies. Therefore, the success of PCR in microfluidic chips significantly depends on the surface chemistry, in contrast with conventional PCR in test tubes. Many studies have been performed on the PCR compatibility of silicon, silicon dioxide and other materials used in microfluidics. Shoffner *et al.* found that native silicon and silicon nitride (Si_3N_4) were inhibitors of PCR and the SiO_2 surface gave comparable amplifications. [10] Potrich *et al.* analyzed the bio-compatibility of different silicon-based materials by atomic force microscopy (AFM), X-ray photoelectron spectroscopy (XPS) and fluorescence microscopy. [11] Panaro *et al.* evaluated and identified several common plastics, plastic tubing, and disposable syringes which were compatible with the PCR reaction. [12] Prakash *et al.* systematically studied Taq polymerase adsorption on 13 kinds of materials in microfluidic devices. [13] However, a comprehensive and conclusive result is still missing from the above studies.

The adsorption of bio-active components can be intensively investigated by delicate protein–surface interaction measurements, in which bulky and expensive facilities are required, such as AFM, scanning or transmission electron microscopy (SEM or TEM), spectrophotometric protein concentration measurement, Fourier transform infrared spectroscopy (FTIR) or XPS. Moreover, a strict experimental environment is also required in some tests, e.g., SEM, TEM and XPS must be performed in vacuum.

These techniques are inconvenient for the evaluation on a large number of microfluidic materials; therefore we designed a relatively quick measurement simply based on PCR.

In contrast with conventional three-step or temperature PCR, including denaturing, annealing and extension, we simplified PCR into two-temperature by merging annealing and extension steps, and achieved rapid amplifications of the male-specific sex determining region Y (SRY) gene marker by utilizing raw saliva in both conventional PCR and micro PCR devices. The optimized PCR contained the following components: 0.75 μM primers (Life Technologies Co., Carlsbad, CA), 3.5 mM $MgCl_2$ (Kapa Biosystems, Cambridge, MA), 0.2 mM dNTP (Takara Bio Inc., Otsu, Japan), 1× fast buffer I reaction buffer, up to 0.2 mM (0.008%) cresol red (Sigma-Aldrich, Co., St. Louis, MO), 1.2 M betaine (Sigma), 2×10^6 pEYFP-C1 template molecules, and 0.025 U/μl SpeedStar HS DNA polymerase (Takara Bio Inc.). The optimized temperatures for denaturing and annealing for designed primer pair were 71 °C and 91 °C respectively. [14]

PCR Setting and Material Selection

Some solutions for the surface passivation by passive or active coating that avoid the material inhibitory effects in enzymatic reactions have been reported. The additive BSA was also reported to compete with DNA polymerase for the adsorption by the chip surface to improve PCR productivity. [8,15] Also, adsorption of BSA by both hydrophilic and hydrophobic surfaces has been widely examined. [16] Therefore, we applied BSA into PCR components to assess the material inhibitory properties. Additionally, BSA also acts as a polymerase competitor in the inhibitor chelation, [17] facilitates the primer to bind to the template, and stabilizes DNA and polymerase. The reported ideal BSA concentration in PCR ranges from 0.5 to 1.0 μg/μl, somehow, BSA has been used in some reported microchips at the concentration of 2.5 μg/μl. In fact, we found that BSA at the concentration of 2 μg/μl had no inhibitory influence on PCR.

In order to screen PCR-compatible microfluidic materials, we selected a wide range of materials, including polymethyl-methacrylate (PMMA) (cast acrylic sheets Clarex A from

Nitto Jushi Kogyo Co. Ltd.), polycarbonate (PC), polyvinyl chloride (PVC), polypropylene (PP, derived from a 200 μl PCR tube), polytetrafluoroethylene (PTFE), cured polydimethylsiloxane (PDMS), three kinds of wax with melting temperatures (T_m) of 56 °C (white wax: paraffin from Nacalai Tesque), 60 °C (yellow wax: shiftwax from Nikka Seiko) and 80 °C (black wax: wax W from Apiezon), silicon (raw silicon wafer), SiO_2 of 560 nm thickness (carrier silicon wafer), quartz, pyrex glass, indium tin oxide (ITO) glass, soda-lime glass, cured SU-8 epoxy-based negative photoresist (SU-8), Norland Optical Adhesives 61 (NOA61) and 68 (NOA68) exposed to UV light for 2 min, dry epoxy and acrylic glues, metallic iron tubes with the diameter of 0.9 mm and 12.9 mm in length, and mineral oil for molecular biology use (Sigma). Then these selected tiny material fragments or liquids were immersed into PCR reactions and incubated on ice for 30 min before reaction on thermocycler.

After the reaction was completed, detection of PCR products was achieved by 4% agarose gel analysis containing SYBR Safe DNA stain (Life Technologies) and subsequent gel imaging. The relative amplified DNA amounts from different testing material-contained PCR mixture were quantified by measuring relative band intensity (RBI) from the gel images using ImageJ version 1.43 software (developed at the National Institutes of Health).

In contrast to the control containing no testing materials, RBI values revealed which materials are PCR-inhibitory or PCR-friendly. High RBI values indicated successful PCR as DNA was amplified, and therefore the material was more PCR-compatible (less inhibitory).

PCR RESULTS AND DISCUSSION

As shown in Table I, the PCR-friendly materials, exhibiting similar BRI values to that of the control in the PCR outcome, are PP, PTFE, PDMS, wax (T_m 80 °C), SiO_2 quartz, pyrex and soda-lime glasses, NOA68, and mineral oil. The addition of BSA did not improve PCR productivity of these PCR-friendly materials. In contrast, PMMA, waxes, ITO glass, SU-8, NOA61, epoxy and acrylic glues showed complete inhibitory effects on PCR outcome, since they gave extremely low RBI values (RBI $< 1.9 \times 10^2$), indicating almost no amplified DNA. Some other partial PCR-inhibitory materials, including PC, PVC, silicon, SiO_2 coated silicon and metal tubes, showed a lower RBI value (RBI $< 8.4 \times 10^2$) than that of the control. However, the addition of BSA in PCR mixture resulted in significant strong gel band signals (RBI $> 1.4 \times 10^3$) for PMMA, PC, PVC, wax (T_m 56 °C), silicon, SiO_2 coated silicon, ITO glass, acrylic glues, NOA61, and the metal tubes. However, the PCR-inhibitory effects of wax (T_m 60 °C), SU-8 and the epoxy glue could not be rescued by the additive BSA. The failure of PCR has most likely resulted from the inhibition or adsorption of PCR components by the added PCR-inhibitory materials. Compared with other components with smaller molecular weight and less biological activity, the key components of the PCR mixture, DNA and the polymerase, are more prone to inhibition. Therefore, we next determined which key component, DNA or the polymerase (or even both), was adsorbed or inhibited by those additive PCR-inhibitory materials (as shown in Figure 1). Briefly, two different kinds of PCR mixtures were prepared with either the absence of template DNA or the polymerase. After distribution into tubes containing material fragments and incubation on ice for 30 min, the PCR mixtures were removed from the material into the new tubes. Then either the SpeedStar polymerase or template DNA was added to the PCR mixture lacking the polymerase or template DNA respectively. Finally, PCR was performed and the outcome was examined as described above.

Table I. PCR inhibition through material interaction with template DNA or the polymerase correspondingly. The measured RBI is indicated

Material	RBI, interaction with DNA	RBI, interaction with polymerase
PMMA	4.6×10^3	3.4×10^2
PC	4.0×10^3	1.5×10^3
PVC	5.4×10^3	9.1×10^2
PP	4.1×10^3	1.7×10^3
PTFE	3.5×10^3	1.6×10^3
PDMS	4.5×10^3	1.7×10^3
Wax (T_m 56 °C)	3.7×10^3	3.2×10^2
Wax (T_m 60 °C)	9.2×10^2	1.0×10^2
Wax (T_m 80 °C)	4.3×10^3	9.2×10^2
Silicon	3.4×10^3	4.3×10^2
SiO_2 5600 Å	3.8×10^3	1.5×10^3
SiO_2 quartz	3.7×10^3	1.1×10^3
Pyrex glass	3.3×10^3	2.0×10^3
ITO glass	4.2×10^3	1.8×10^2
Soda-lime glass	4.3×10^3	1.7×10^3
SU8	4.1×10^3	1.0×10^2
NOA61	1.5×10^3	8.1×10^1
NOA68	3.9×10^3	3.9×10^3
Epoxy glue	4.2×10^3	8.0×10^1
Acrylic glue	9.4×10^2	8.4×10^1
Metal tubes	3.0×10^3	1.5×10^2
Mineral oil	7.6×10^2	3.4×10^1
No additives	9.8×10^2	2.0×10^2

The RBI value showed that the wax (T_m 60 °C) exhibited complete adsorption effect on template DNA (RBI = 9.2×10^1), NOA61, mineral oil and acrylic glue materials significant adsorption occurred (RBI < 1.5×10^3), but the other examined materials did not noticeably adsorb the template DNA (RBI > 3.0×10^3). This is consistent with the literature report that DNA is not expected to bind to hydrophobic surfaces due to its poly-anionic feature. In contrast to DNA, we observed more significant effects of polymerase adsorption on the PCR efficiency. The following materials did not have strong effects (RBI > 1.1×10^3) on polymerase: PC, PP, PTFE, PDMS, silicon with a layer of 560 nm SiO_2, SiO_2 quartz, pyrex, and soda-lime glass. Slight polymerase inhibition (RBI < 9.2×10^2) was observed with PMMA, PVC, waxes (T_m 56 °C and 80 °C), silicon, and NOA68. A very strong or near total inhibition (RBI < 1.8×10^2) was observed with wax (T_m 60 °C), ITO glass, SU-8, NOA61, metal tubes, mineral oil, epoxy, and the acrylic glues.

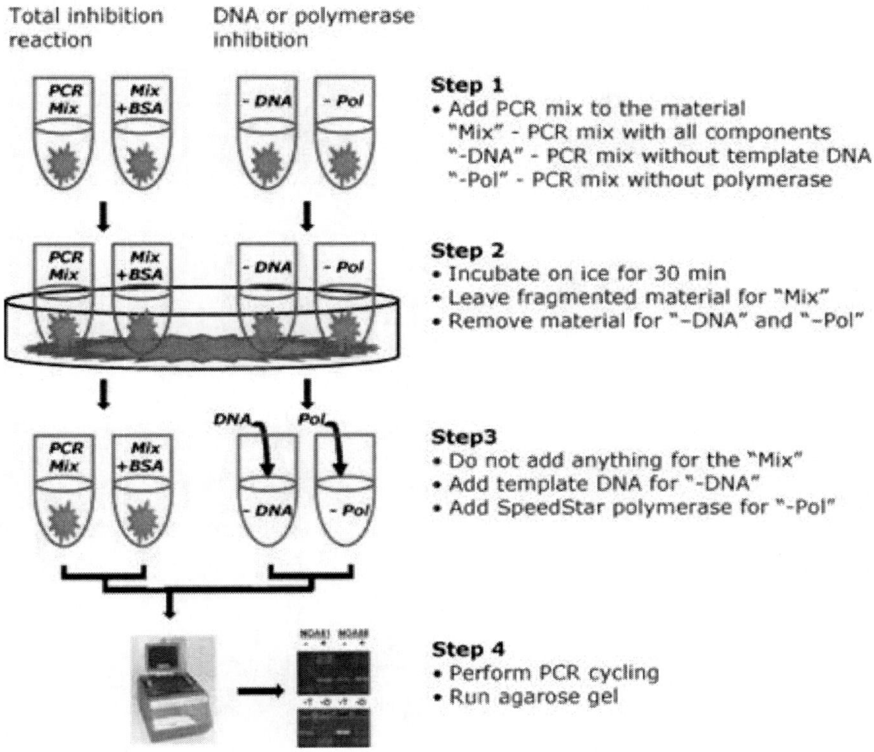

Figure 1. The PCR compatibility assay. Both the total reaction inhibition experiment and DNA or polymerase adsorption experiments are described. The PCR is performed using bench thermocycler, and subsequent imaging is done on agarose gel. The control experiment was done running PCR without including any of the material ("no additives" part).). Reprinted with permission from [46]. Copyright 2012, Elsevier.

MATERIAL INHIBITION EFFECT

As summarized in Table I and II, we totally examined nine major categories of materials commonly used in microfluidic devices or PCR applications, i.e., silicon-based materials, oxide, photoresist, photopolymer, metals, mineral oil, plastics, wax and glue. These materials are in either solid or liquid form, resulting in varied shapes and total surface areas, which range from 3.5×10^1 to 1.8×10^2 mm^2. And the surface-area-to-volume ratio (SAVR) of these materials, which may dramatically influence reaction outcomes in microfluidic devices, may not be at the same comparable level. Therefore, the materials listed in Table I should be compared with caution and more delicate and precise sample preparation measures should be employed in future work to maintain the similar SAVR of each material. The PCR-compatibility of silicon-based microdevices has been examined by many researchers. Our finding is consistent with the literature that bare silicon (RBI = 2.2×10^2) has stronger inhibitory effect than SiO_2 (RBI = 4.4×10^2). And we clarified that this inhibitory effect was exerted on the polymerase but not the template DNA and could be rescued by additive BSA. Interestingly, quartz, which is crystalline form of SiO_2 and has been used in microfluidics for single cell analysis, lipoprotein assay, cell separation, cell counting and biosensor, [18-22] did not show any inhibitory effect on DNA or the polymerase. The reason is not clear and needs further investigation.

Table II. Summary on PCR inhibitory effect of various microfluidic materials according to the calculated RBI in Table I

Category	Material	Inhibition	BSA rescue	
		DNA	Polymerase	
Silicon	Silicon	Insignificant	Slight	Yes
	SiO_2	Insignificant	Insignificant	Yes
	Quartz	Insignificant	Insignificant	No
Oxide glass	Pyrex	Insignificant	Insignificant	No
	Soda-lime	Insignificant	Insignificant	No
	ITO	Insignificant	Very strong	Yes
Photoresist	SU8	Insignificant	Very strong	No
Photopolymer	NOA61	Significant	Very strong	Yes
	NOA68	Insignificant	Slight	No
Metal	Metal tube	Insignificant	Very strong	Yes
Mineral oil	Mineral oil	Significant	Very strong	No
Plastics	PP	Insignificant	Insignificant	No
	PTFE	Insignificant	Insignificant	No
	PDMS	Insignificant	Insignificant	No
	PMMA	Insignificant	Slight	Yes
	PC	Insignificant	Insignificant	Yes
	PVC	Insignificant	Slight	Yes
Wax	Wax (T_m 56°C)	Insignificant	Slight	Yes
	Wax (T_m 60°C)	Very strong	Very strong	No
	Wax (T_m 80°C)	Insignificant	Slight	No
Glue	Acrylic glue	Significant	Very strong	Yes
	Epoxy	Insignificant	Very strong	No

We also examined three kinds of oxide, i.e., pyrex, soda-lime, and ITO, which have been popular for glass microfluidic devices in cell culture, separation and real-time cellular imaging systems. [33-36] The ITO glass (RBI = 9.1×10^1) exhibited the strong inhibitory effect on the PCR outcome that could be rescued by the additive BSA.

This inhibitory effect was demonstrated to affect the polymerase but not the template DNA. In contrast, both pyrex and soda-lime glasses (RBI = 3.3×10^3 and 2.8×10^3, respectively) did not inhibit either DNA or the polymerase to affect PCR outcome, which is consistent with literature report. [11]

SU-8, as a photoresist, and NOA 61 and NOA68, as the liquid photopolymers, have been commonly used in photolithography. SU-8 (RBI < 8.2×10^1) was found to strongly inhibit the polymerase but not DNA, leading to poor amplification, and the PCR outcome could not be improved by the additive BSA. This result is consistent with previous reports. [23] NOA61 and NOV68 are different in physical features and usage. NOA68 has a higher viscosity (22,000 CPS) than NOA61 (300 CPS). NOA61 serves as an adhesive for glass and metal, while NOA68 is used to adhere to the plastic.

Our results suggest that NOA68 (RBI = 2.9×10^3) is much more PCR-friendly than NOA61 (RBI = 1.9×10^2) and the inhibitory effect of NOA61 on both DNA and the polymerase could be rescued by the additive BSA.

Acrylic glue and epoxy glue have been used for chip bonding. We found that both acrylic glue (RBI=6.0×10^1) and epoxy glue (RBI= 9.1×10^1) were PCR-inhibitory. Acrylic glue could inhibit both DNA and the polymerase significantly while epoxy glue appeared to inhibit the polymerase but not DNA. However, acrylic glue turned out to be more PCR-friendly with the additive BSA (RBI=5.9×10^2), whereas the PCR with the epoxy glue still failed in the presence of BSA (RBI = 1.1×10^2). Therefore, acrylic glue will be recommended for use in PCR-compatible microdevices instead of epoxy glue.

The stainless metal tubes are used as physically stable connections between tubing and microfluidic devices. Usually the metal tubes are beyond the contact with the reaction vessel of PCR microfluidic chips, so little attention has been paid to their influence on PCR. Our results showed that the metal tubes reduced the PCR yield through polymerase inhibition rather than template DNA, which is consistent with Panaro's work. [12] Other metals such as platinum and gold have also been reported to inhibit PCR and the suggested solution to this inhibition was the application of EDTA with the equal amount of Mg^{2+}. [24-26] The mineral oil purchased from Sigma-Aldrich is free of DNase, RNase and Protease and suitable for molecular biology usage. Due to lower density and higher boiling temperature than water and inertness, this kind of mineral oil has been used to overlay PCR mixture to maintain the constant concentration of components by preventing water evaporation, depending on the model of thermocycler. We found that this mineral oil indeed had little influence on PCR yield compared with control, which confirms the PCR-compability of this commercial mineral oil.

The bio-compatibility of several plastics has been evaluated by several research groups. Panaro *et al.* found that native acrylic, Kel-F® (chlorotrifluoroethylene), polysulfone and PTFE Teflon® did not inhibit PCR, whereas native Delrin® and poly(etheretherketone) (PEEK) reduced PCR yield significantly. [12] Midwoud *et al.* assessed several plastic microfluidic materials as alternatives for PDMS devices, including PMMA, polystyrene (PS), PC, and cyclic olefin copolymer (COC). They found that only UV-ozone-treated PC and COC devices satisfied bio-compatibility on cell culture and adsorption of hydrophobic drugs and their metabolites. [27] The six different plastics we examined in this publication have widely been applied in the fabrication of conventional tools and microfluidic chips. [24,28-31] We found that all plastic materials showed unnoticeable or slight PCR inhibitory effect and only PMMA and PVC slightly inhibit the polymerase but not DNA. PP, PTFE, and PDMS had the least inhibitory effect on PCR and the additive BSA could rescue the PCR inhibitory effect of PMMA, PC, and PVC. Our results suggested that these plastic materials can be used for PCR-compatible microfluidic devices with the additive BSA. Waxes are a class of non-homogeneous chemical compounds that are malleable near ambient temperatures. All waxes are water-resistant and composed of various substances, including long-chain carbons (from 12 to 38 carbon atoms), hydrocarbons, ketones, alcohols, aldehydes, sterol esters, alkanoic acids, terpenes, and monoesters. The melting point of waxes can be as low as 45 °C so wax can influence PCR yield by its chemical components as well as some by-products generated at high temperature. As recently wax has been utilized in the fast fabrication of simple microfluidic devices, [32-40] we tested three available kinds of waxes with different melting points. We found that wax (T_m 80 °C) exhibited good PCR-compatibility even without the additive BSA. Also, the PCR-compatibility of wax (T_m 56 °C) could be significantly enhanced by the additive BSA (RBI = 1.5×10^3). Both wax (T_m 80 °C) and wax (T_m 56 °C) slightly inhibit polymerase rather than template DNA. Wax (T_m 60 °C), however, completely inhibited the reaction despite the additive BSA (RBI = 1.9×10^2). Wax (T_m 60 °C) showed very strong inhibition on both DNA template and the polymerase; the PCR could be rescued by doubling the polymerase concentration as reported in Gong *et al.* [32] In summary, we introduced a simple method to access the PCR-compatibility of various microfluidic materials, which can be easily performed in common molecular biology laboratories and which enables us to compare among a much wider range of materials than previous studies. Our study has provided a comprehensive insight into the selection of PCR-friendly microfluidic materials. Furthermore, other biological studies can also benefit from our strategy, such as cell growth inhibition study, [41] high-throughput screening and transcriptome analysis. In addition, based on our compatibility test results of the waxes and glues, we are able to introduce novel materials and fabrication methods for microfluidics, such as mask-free and paper-based two-dimensional (2D) and three-dimensional (3D) chips.

MASK-FREE AND PAPER BASED MICROFLUIDIC DEVICE FOR BIOLOGICAL ANALYSIS

Soft-lithography, using a soft polymer such as PDMS to imprint and transfer the structure on a well patterned mold, is the most popular and successful microfluidic fabrication method. However, the whole fabrication process is very time-consuming, requiring several hours or even days to deal with mask making and photolithography. Also, it is difficult to construct complex 3D structures by multilayer bonding. Therefore, in order to improve the fabrication efficiency and develop new applications, many rapid and inexpensive fabrication methods have been introduced. One of these methods is paper-based microfluidics.

As it is widely available, low-cost and compatible with many chemical and biological applications, paper-based microfluidics has provided a novel system for microfluidic manipulation ever since 2007. [14,32] Currently, paper-based microfluidic systems have shed new light on disease diagnosis and environmental monitoring for those areas especially lacking equipment and professionals. A variety of 2D and 3D microfluidic devices have been reported regarding the fabrication techniques or/and applications. For instance, Delaney *et al.* succeeded in combining paper microfluidics with electrochemiluminescent (ECL) detection which can be recorded by a mobile phone camera instead of a traditional photodetector. [42] Ho *et al.* successfully detected drugs at nanomolar concentrations by coupling a surface acoustic wave atomization and ionization device with a paper-based sample delivery system and mass spectrometer. [43] Dungchai *et al.* firstly demonstrated the electrochemical determination of glucose, lactate, and uric acid in biological samples using oxidase enzyme reactions by paper-based microfluidic devices. [44] Dungchai *et al.* also reported a rapid fabrication method for paper-based microfluidics using wax screen-printing. [45]

In this publication, we report a one-step gas-sacrificing glue bonding method to fabricate microfluidic systems for biological analysis. In this process, instead of tedious soft-lithography, a CO_2 laser system (Versa Laser System, model VR3.50, Universal Laser Systems, Ltd., Scottsdale, AZ) was used to cut through paper to form intricate patterns and configured channels. The cyanoacrylate-based resin (Aron Alpha Co.,West Jefferson, OH) is a rapid-polymerizing liquid glue in medical use which is compatible for various kinds of material surface and offers excellent bonding strength. By bonded with cyanoacrylate-based resin, the laser-cut paper sheets were sandwiched between hydrophilic glass slides or hydrophobic polymer-based plates to form a multilayer structure. The whole process could be finished within 1 hour and the cost of materials was less than 1 US Dollar. With this simple, rapid and low-cost method, not only intricate hydrophilic or hydrophobic channels can be easily fabricated, but 3D structures have also been realized by one-step bonding of multiple layers (as shown in Figure 2). [14]

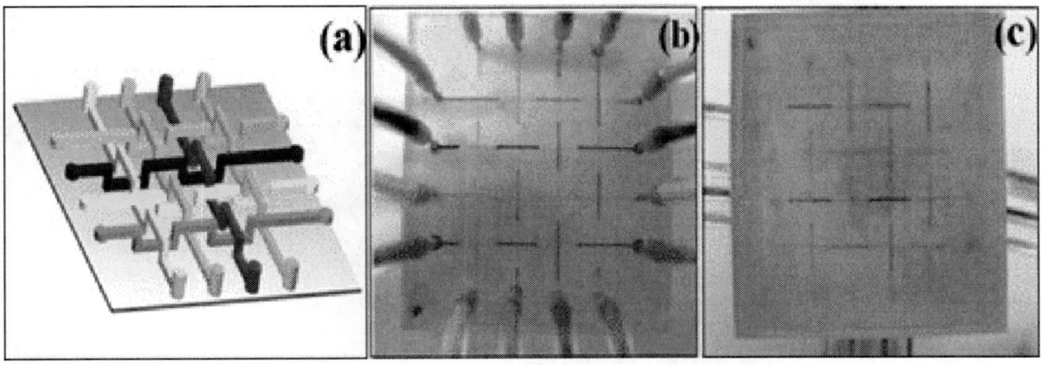

Figure 2. (a) Schematic picture of three-dimensional chip. Differently colored channels intersected and overlapped with each other without connecting. (b) Top-view of three-dimensional acrylic resin microfluidic chip. Each channel was filled with one color of fluid, corresponding to (a). (c) Bottom view of three-dimensional acrylic resin microfluidic chip. Reprinted with permission from [14]. Copyright 2010, AIP Biomicrofluidics.

We then evaluated the PCR-compatibility of materials (A4 print paper, filter paper, and cyanoacrylate-based resin) employed in the chip fabrication by the method described above. The results showed that resin impregnated paper could avoid the PCR-inhibitory effect of pure paper. Thus this 2D and 3D microfluidic chips were bio-compatible with the enzymatic reactions. Protein detection and DNA capillary electrophoresis have also been carried out and both of them showed positive results. Overall, with the direct laser cutting and one-step gas-sacrificing techniques, the whole fabrication process for complicated 2D and 3D microfluidic devices has successfully been shortened into several minutes, making it a potential alternative of PDMS microfluidic chips for biological analysis.

EASILY ACHIEVED WAX-BONDING 3D MICROFLUIDIC CHIPS

We also fabricated another simple, low-cost and detachable microfluidic chip by bonding paper, glass slides or other polymer films with adhesive wax (Nikka Seiko (Skywax Series, Japan)). We firstly used the scalpel or the same CO_2 laser system as described above to cut through the paper or film to form patterns and then sandwich the paper or film between glass sheets or polymer membranes. Then, the paper or film absorbed the hot melt adhesive wax and the cooled wax-paper or wax-film was sandwiched between two glass slides. With this process, 3D microfluidic chip is achievable by vacuating and venting the chip in a hot-water bath.

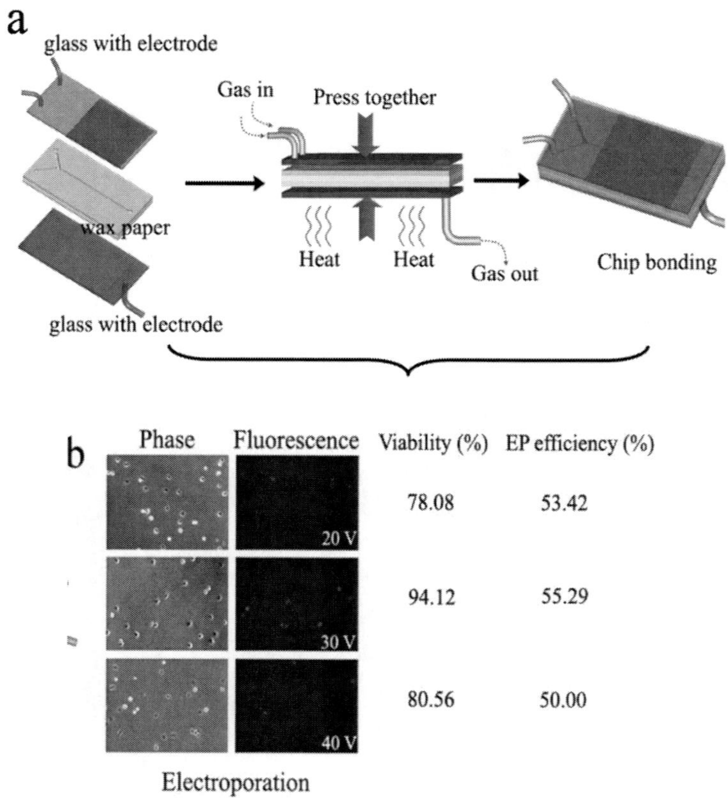

Figure 3. (a) The flow chart to show the fabricating process of a wax-paper-based microfluidic device with imbedded electrodes. (b) The phase contrast and fluorescent images of *HeLa* cells after EP in the chip described in (a). The cell viability and the EP efficiency were obtained from cell morphology in phase contrast images and the percentage of Rhodamin-Dextran positive *HeLa* cells respectively. Reprinted with permission from [32]. Copyright 2010, RSC Publishing.

By SEM scanning the surface morphology, we found that the bonding strength of wax chips could be sustained and wax-film gave a smoother surface than wax-paper. Also, chips based on the transparent film enabled the optical analysis that was not possible with the paper-based chips.

As shown in Figure 3, we fabricated a 5-layer 3D metal electrodes incorporated chip by multilayer wax bonding and then applied it for the electroporation (EP) of *HeLa* cells (ATCC, Manassas, VA). No liquid leakage could be seen in the channel even after repetitive use, suggesting that the wax bonding provided a stable sealed compartment. Also, no cells were trapped under microscope observation in wax-based chip, suggesting the excellent cell-compatibility of wax. The *HeLa* cells were suspended in the poration medium containing dextran, rhodamine B isothiocyanate, and 70S (Sigma), serving as fluorescent EP marker. The electrodes were connected to a radio-frequency (RF)-oscillating electric pulse generator (Bio-Rad Labs). After applying different voltages, the cells were analyzed under bright and fluorescent field. The mean viability of *HeLa* cells after electroporation is 84.25 ± 8.63%, accompanying with the EP efficiency 52.91 ± 2.68%, which is as good as the conventional EP method. The wax-bonded electrodes can also be applied to other electro-dynamic processes, e.g., electrophoresis and electro-detection. Moreover, the 5-layer 3D device offers good optical transparency and low background fluorescence, which allows optical studies on cell culture and drug screening. [32]

CONCLUSION

In this publication, by using a simple and rapid PCR based approach, we assessed extensively nine common categories of materials used in microfluidic devices. Our approach established the simple and applicable concept for the further screening of mass amounts of prospective microfluidic materials.

Moreover, our conclusions can serve as a good guidance for the selections of materials for the fabrication of microfluidic devices. [46] We also demonstrated two examples of utilizing the simple, low-cost and bio-compatible materials and fabrication methods in microfluidic devices, which may help researchers to develop more applicable microdevices for rapid on-site diagnosis, monitoring and quality control.

ACKNOWLEDGMENTS

This publication is based on work supported by Seed Fund Award "Envirola" made by King Abdullah University of Science and Technology (KAUST).

REFERENCES

[1] Chokkalingam, V. *et al.* Optimized droplet-based microfluidics scheme for sol-gel reactions. *Lab on a chip* 10, 1700-1705, doi:10.1039/b926976b (2010).

[2] Whitesides, G. M. The origins and the future of microfluidics. *Nature* 442, 368-373, doi:10.1038/nature05058 (2006).

[3] Domachuk, P., Tsioris, K., Omenetto, F. G. and Kaplan, D. L. Bio-microfluidics: biomaterials and biomimetic designs. *Advanced materials* 22, 249-260, doi:10.1002/adma.200900821 (2010).

[4] Jaeger, R. C. *Introduction to Microelectronic Fabrication, Second Edition*. 2 edn, 316 (Prentice Hall, 2002).

[5] Saiki, R. K. *et al.* Enzymatic amplification of beta-globin genomic sequences and restriction site analysis for diagnosis of sickle cell anemia. *Science* 230, 1350-1354 (1985).

[6] deMello, A. J. Microfluidics: DNA amplification moves on. *Nature* 422, 28-29, doi:10.1038/422028a (2003).

[7] Wu, J., Kodzius, R., Xiao, K., Qin, J. and Wen, W. Fast detection of genetic information by an optimized PCR in an interchangeable chip. *Biomedical microdevices* 14, 179-186, doi:10.1007/s10544-011-9595-6 (2012).

[8] Zhang, C., Xu, J., Ma, W. and Zheng, W. PCR microfluidic devices for DNA amplification. *Biotechnology advances* 24, 243-284, doi:10.1016/j.biotechadv. 2005. 10.002 (2006).

[9] Lagally, E. T., Emrich, C. A. and Mathies, R. A. Fully integrated PCR-capillary electrophoresis microsystem for DNA analysis. *Lab on a chip* 1, 102-107, doi:10.1039/b109031n (2001).

[10] Shoffner, M. A., Cheng, J., Hvichia, G. E., Kricka, L. J. and Wilding, P. Chip PCR. I. Surface passivation of microfabricated silicon-glass chips for PCR. *Nucleic acids research* 24, 375-379 (1996).

[11] Potrich, C. *et al.* Effect of materials for micro-electro-mechanical systems on PCR yield. *European biophysics journal : EBJ* 39, 979-986, doi:10.1007/s00249-009-0466-5 (2010).

[12] Panaro, N. J., Lou, X. J., Fortina, P., Kricka, L. J. and Wilding, P. Surface effects on PCR reactions in multichip microfluidic platforms. *Biomedical microdevices* 6, 75-80 (2004).

[13] A. Ranjit Prakash, M. A. a. K. V. I. S. K. Characteristics and impact of Taq enzyme adsorption on surfaces in microfluidic devices. *Microfluidics and nanofluidics* 4, pp 295-305, doi:10.1007/s10404-007-0174-3 (2008).

[14] Yi, X., Kodzius, R., Gong, X., Xiao, K. and Wen, W. A simple method of fabricating mask-free microfluidic devices for biological analysis. *Biomicrofluidics* 4, doi:10.1063/1.3487796 (2010).

[15] Zhang, C. and Xing, D. Miniaturized PCR chips for nucleic acid amplification and analysis: latest advances and future trends. *Nucleic acids research* 35, 4223-4237, doi:10.1093/nar/gkm389 (2007).

[16] Taylor, T. B., Winn-Deen, E. S., Picozza, E., Woudenberg, T. M. and Albin, M. Optimization of the performance of the polymerase chain reaction in silicon-based microstructures. *Nucleic acids research* 25, 3164-3168 (1997).

[17] Kreader, C. A. Relief of amplification inhibition in PCR with bovine serum albumin or T4 gene 32 protein. *Applied and environmental microbiology* 62, 1102-1106 (1996).

[18] Zhuang, G. *et al.* A low temperature bonding of quartz microfluidic chip for serum lipoproteins analysis. *Biomedical microdevices* 8, 255-261, doi:10.1007/s10544-006-9142-z (2006).

[19] Greif, D., Galla, L., Ros, A. and Anselmetti, D. Single cell analysis in full body quartz glass chips with native UV laser-induced fluorescence detection. *Journal of chromatography. A* 1206, 83-88, doi:10.1016/j.chroma.2008.07.013 (2008).

[20] Wang, H. *et al.* Microchip-based small, dense low-density lipoproteins assay for coronary heart disease risk assessment. *Electrophoresis* 29, 1932-1941, doi:10.1002/elps.200700631 (2008).

[21] Kim, D. J. *et al.* A quartz nanopillar hemocytometer for high-yield separation and counting of CD4(+) T lymphocytes. *Nanoscale* 4, 2500-2507, doi:10.1039/c2nr11338d (2012).

[22] Lederer, T., Stehrer, B. P., Bauer, S., Jakoby, B. and Hilber, W. Utilizing a high fundamental frequency quartz crystal resonator as a biosensor in a digital microfluidic platform. *Sensors and actuators. A, Physical* 172, 161-168, doi:10.1016/j.sna.2011.04.032 (2011).

[23] T.B. Christensen, C. M. P., T.G. Grondahl, T.G. Jensen, A. Sekulovic, D.D. Bang, A. Wolff. PCR biocompatibility of lab-on-a-chip and MEMS materials. *Journal of Micromechanics and Microengineering* 17 (2007).

[24] Ramalingam, N. *et al.* Real-time PCR array chip with capillary-driven sample loading and reactor sealing for point-of-care applications. *Biomedical microdevices* 11, 1007-1020, doi:10.1007/s10544-009-9318-4 (2009).

[25] Wan, W. and Yeow, J. T. The effects of gold nanoparticles with different sizes on polymerase chain reaction efficiency. *Nanotechnology* 20, 325702, doi:10.1088/0957-4484/20/32/325702 (2009).

[26] Teng, F., Guan, Y. and Zhu, W. A simple and effective method to overcome the inhibition of Fe to PCR. *Journal of microbiological methods* 75, 362-364, doi:10.1016/j.mimet.2008.06.013 (2008).

[27] van Midwoud, P. M., Janse, A., Merema, M. T., Groothuis, G. M. and Verpoorte, E. Comparison of biocompatibility and adsorption properties of different plastics for advanced microfluidic cell and tissue culture models. *Analytical chemistry* 84, 3938-3944, doi:10.1021/ac300771z (2012).

[28] Li, Y., Xing, D. and Zhang, C. Rapid detection of genetically modified organisms on a continuous-flow polymerase chain reaction microfluidics. *Analytical biochemistry* 385, 42-49, doi:10.1016/j.ab.2008.10.028 (2009).

[29] Dahl, A. *et al.* Quantitative PCR based expression analysis on a nanoliter scale using polymer nano-well chips. *Biomedical microdevices* 9, 307-314, doi:10.1007/s10544-006-9034-2 (2007).

[30] Gervais, L. and Delamarche, E. Toward one-step point-of-care immunodiagnostics using capillary-driven microfluidics and PDMS substrates. *Lab on a chip* 9, 3330-3337, doi:10.1039/b906523g (2009).

[31] Zhou, X. *et al.* Microfluidic PicoArray synthesis of oligodeoxynucleotides and simultaneous assembling of multiple DNA sequences. *Nucleic acids research* 32, 5409-5417, doi:10.1093/nar/gkh879 (2004).

[32] Gong, X. *et al.* Wax-bonding 3D microfluidic chips. *Lab on a chip* 10, 2622-2627, doi:10.1039/c004744a (2010).

[33] Songjaroen, T., Dungchai, W., Chailapakul, O. and Laiwattanapaisal, W. Novel, simple and low-cost alternative method for fabrication of paper-based microfluidics by wax dipping. *Talanta* 85, 2587-2593, doi:10.1016/j.talanta.2011.08.024 (2011).

[34] Wang, S. *et al.* Paper-based chemiluminescence ELISA: lab-on-paper based on chitosan modified paper device and wax-screen-printing. *Biosensors and bioelectronics* 31, 212-218, doi:10.1016/j.bios.2011.10.019 (2012).

[35] Nilghaz, A. *et al.* Flexible microfluidic cloth-based analytical devices using a low-cost wax patterning technique. *Lab on a chip* 12, 209-218, doi:10.1039/c1lc20764d (2012).

[36] Diaz-Gonzalez, M. and Baldi, A. Fabrication of bio-functionalized microfluidic structures by low-temperature wax bonding. *Analytical chemistry*, doi:10.1021/ac301512f (2012).

[37] Carrilho, E., Martinez, A. W. and Whitesides, G. M. Understanding wax printing: a simple micropatterning process for paper-based microfluidics. *Analytical chemistry* 81, 7091-7095, doi:10.1021/ac901071p (2009).

[38] Lu, Y., Shi, W., Qin, J. and Lin, B. Fabrication and characterization of paper-based microfluidics prepared in nitrocellulose membrane by wax printing. *Analytical chemistry* 82, 329-335, doi:10.1021/ac9020193 (2010).

[39] Lu, Y., Shi, W., Jiang, L., Qin, J. and Lin, B. Rapid prototyping of paper-based microfluidics with wax for low-cost, portable bioassay. *Electrophoresis* 30, 1497-1500, doi:10.1002/elps.200800563 (2009).

[40] Kaigala, G. V., Ho, S., Penterman, R. and Backhouse, C. J. Rapid prototyping of microfluidic devices with a wax printer. *Lab on a chip* 7, 384-387, doi:10.1039 /b617764f (2007).

[41] Kunzmann, A. *et al.* Toxicology of engineered nanomaterials: focus on biocompatibility, biodistribution and biodegradation. *Biochimica et biophysica acta* 1810, 361-373, doi:10.1016/j.bbagen.2010.04.007 (2011).

[42] Delaney, J. L., Hogan, C. F., Tian, J. and Shen, W. Electrogenerated chemiluminescence detection in paper-based microfluidic sensors. *Analytical chemistry* 83, 1300-1306, doi:10.1021/ac102392t (2011).

[43] Ho, J. *et al.* Paper-based microfluidic surface acoustic wave sample delivery and ionization source for rapid and sensitive ambient mass spectrometry. *Analytical chemistry* 83, 3260-3266, doi:10.1021/ac200380q (2011).

[44] Dungchai, W., Chailapakul, O. and Henry, C. S. Electrochemical detection for paper-based microfluidics. *Analytical chemistry* 81, 5821-5826, doi:10.1021/ac9007573 (2009).

[45] Dungchai, W., Chailapakul, O. and Henry, C. S. A low-cost, simple, and rapid fabrication method for paper-based microfluidics using wax screen-printing. *The Analyst* 136, 77-82, doi:10.1039/c0an00406e (2011).

[46] R. Kodzius, K. Xiao, J. Wu, X. Yi, X. Gong, I. G. Foulds and W. Wen, *Sensors and Actuators B: Chemical* 161 (1), 349-358 (2012).

In: Microfluidics
Editors: Stefano Panzarella and Walter Maroni
ISBN: 978-1-62417-345-5

Chapter 7

ELECTROKINETIC MANIPULATION OF BIOMOLECULES

Kin Fong Lei*
Graduate Institute of Medical Mechatronics,
Department of Mechanical Engineering,
Chang Gung University, Taoyuan, Taiwan

ABSTRACT

Electrokinetics in microfluidic system provides effective molecular manipulation technique in micro/nano domain, which matches the length scale of biomolecules. A lot of demonstrations have been reported and shown the manipulation of different biomolecules using different kinds of electrokinetic phenomena. This chapter reviewed the most commonly used electrokinetic phenomena including DC electrophoresis, AC electrophoresis, DC electroosmosis, AC electroosmosis, electrowetting, and opto-electrically induced electrokinetics. The principle behind and recent developments on the electrokinetic manipulation will be discussed. An updated and systematic in-depth discussion is provided in the field of electrokinetic manipulation.

1. INTRODUCTION

The development of microfluidic technology opens new avenues to realize the miniaturized equipment for biological analysis applications. Microfluidic system, also called "lab-on-chip (LOC)", "bio-chip", or "micro-total-analysis-system (μTAS)", has been rapidly developed from early single channel devices [1] to recent complex analysis systems [2-4]. One of the major objectives of the development of microfluidic system is to miniaturize the analytical equipment in conventional laboratory. Therefore, entire analytical protocol, including sample pretreatment, sample/reagent manipulation, separation, reaction, and detection, can be performed automatically in a miniaturized complex device. Because

* Email: kflei@mail.cgu.edu.tw.

microfluidic system is designed to handle fluids in micro/nano region, the fluidic operations match the length scale of various biomolecules and provide precise control of processing parameters for analytical process.

In most of the analytical protocols, biomolecules are suspended in fluid medium for the biological operation. Conventionally, fluid medium is driven by on-chip mechanical micropumps to achieve biomolecular manipulation in microfluidic systems [5, 6]. Many strategies for on-chip fluid pumping have been reported, for example, piezoelectric [7-10], pneumatic [11-15], electrostatic [16-18], and electromagnetic [19] pumping. Although these mechanical micropumps were effective for the bulk fluid pumping, but complicated structure and fabrication process involved limited the integration with other processes, such as sample preparation, reaction, and detection, in a single device. Therefore, non-mechanical pumping mechanism was explored for alternative method. The use of electrokinetics is growing interest to manipulate fluids and particles within the microfluidic system using integrated electrodes. The electrokinetic forces are easy to control by designing electrode pattern and the applied potential and frequency. Also, large electric fields can be generated for manipulation using relatively low potentials in microfluidic systems. Moreover, the electrodes can be fabricated in micro/nano scale to match the length scale of various biomolecules for manipulation. In this chapter, recent developments on electrokinetic manipulation of biomolecules will be reviewed. Electrokinetics of DC and AC fields, including DC electrophoresis, AC electrophoresis, DC electroosmosis, AC electroosmosis, electrowetting, and opto-electrically induced electrokinetics are focused. The experimental achievements of biomolecular manipulation in microfluidic systems will be discussed.

2. DC Electrophoresis

DC electrophoresis (or electrophoresis) is a technique to manipulate charged particles suspended in stationary fluid medium under a uniform electric field. The particles move in bulk fluid towards either cathode or anode depending on the sign of the surface charge of the particles. This technique is widely used in biochemical laboratory to separate a mixed population of proteins [20-21] or DNA fragments [22]. The movement of a charged particle suspended in bulk fluid is based on the resultant force of the electrostatic Coulomb force, electrophoretic retardation force, and drag force. Illustration of the electrophoretic motion is shown in Figure 1. The electrostatic Coulomb force is generated by the surface charges of the particle. The electrophoretic retardation force is generated by the electrical double layer (EDL) of the particle, which is two parallel layers of charges surrounding the particle [23]. The drag force is generated by the friction between moving particle and fluid medium. When a particle is suspended in the fluid medium, it is electroneutral due to the presence of the EDL. The charges in the EDL are exactly equal and opposite to the charges on the particle. In this situation, the particle appears to have zero net charge and does not move. But when a uniform electric field is applied across the channel, the particle moves because the counter ions in the EDL are not fixed to the surface of the particle. The counter ions have the same mobility as the bulk solution and move in the direction opposite to the particle. Since the particle is very small and the fluid is viscous in the micro/nano scale, the global scale of the fluid is stationary. Electrophoretic motion of rigid particles has been investigated mathematically to describe this phenomenon in detail [23]. This section will focus on the discussion of the experimental achievement of using electrophoresis to manipulate biomolecules.

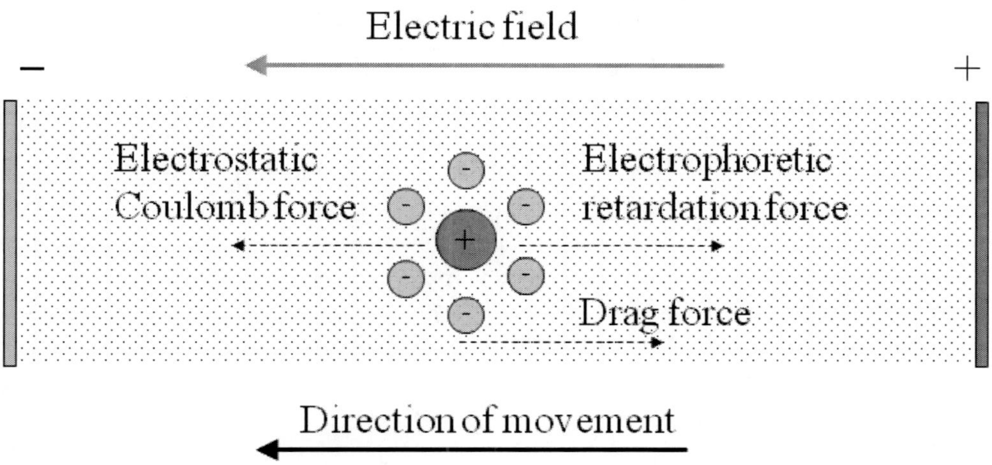

Figure 1. Illustration of the electrophoretic motion.

Biomolecules generally have some surface charges and can be manipulated by electrophoretic force. Due to different molecular weights of the biomolecules, they exhibit different electrophoretic mobilities and can therefore be separated by electrophoresis in capillary tube. In the past decade, capillary electrophoresis (CE) devices have been reported extensively and used to separate biomolecules including amino acids, viruses, and proteins [24-26]. The CE devices normally were incorporated with detection mechanisms, such as fluorescence, absorbance, electrochemical, and refractive index [27, 28]. Therefore, biomolecules can be manipulated and analyzed in the CE devices simutaneously. Baldwin et al. developed a CE device integrated with electrochemical detection [29]. Both CE channels and electrodes were incorporated directly onto glass substrates fabricated by standard microfabrication techniques, including photolithographic patterning, wet chemical etching, DC sputtering, and thermal wafer bonding. The layout of the CE device is shown in Figure 2. Critical electrodes for manipulating biomolecules and electrochemical detection were fabricated onto the bottom glass substrate in order to eliminate the need for external electrodes. CE channels (20 μm deep and 50 μm wide) and four reservoir openings (sample, waste, buffer, and detection) were fabricated on the top glass substrate. The top and bottom substrates were cleaned and then bonded by thermal bonding. Manipulation of the biomolecules was based on the control of the application of high voltages on the electrodes. The electrophoresis procedure of the CE device is described and this is also the general procedure for most of the CE device. To begin the experiment, all channels were filled with the running buffer. Then, the buffer in the sample reservoir was emptied and replaced by the sample solution. A high voltage was applied to the electrode in the waste reservoir while electrodes in the sample, buffer, and detection reservoirs were all grounded. Sample was manipulated along the injection channel and the intersection volume was then filled with sample. To start the separation, a high voltage was applied to the electrode in the buffer reservoir and all other electrodes were held at ground. Electrophoresis was started to separate the sample from the intersection volume to the detection reservoir for electrochemical analysis. Here, to achieve the 250 V/cm electric fields required for such short-length CE channels, Baldwin et al. used a battery-operated power supply delivering 250 V and 2000 V DC for the injection and separation channels, respectively.

CE becomes an important tool for analysis of various biomolecules. For example, silicon polymerase chain reaction (PCR) reactor and glass CE chip have been successfully coupled to form an integrated DNA analysis system [30]. The PCR reactor provided a rapid thermal cycling capability (10 °C/s heating and 2.5 °C/s cooling) and the CE chip can perform high-speed (120 s) DNA separation. The functionality was demonstrated by PCR-CE analysis of β-globin in 20 min and *Salmonella* DNA in 45 min. This integrated DNA analysis system showed the capability of fabrication of complex and integrated microsystem for real-time monitoring of PCR amplification. Moreover, a multilayer polymer CE device for protein analysis has been reported for applications in biology, chemical analysis, drug discovery, and disease screening [31]. It was desirable to have inexpensive and high-throughput system that integrates multiple sample analysis processes and procedures. This work demonstrated a multilayer polymer microfluidic device integrated with on-chip labeling and parallel electrophoretic separation of up to eight samples. Individual proteins and a mixture of cancer biomarkers have been successfully labeled on-chip and separated in parallel with this system.

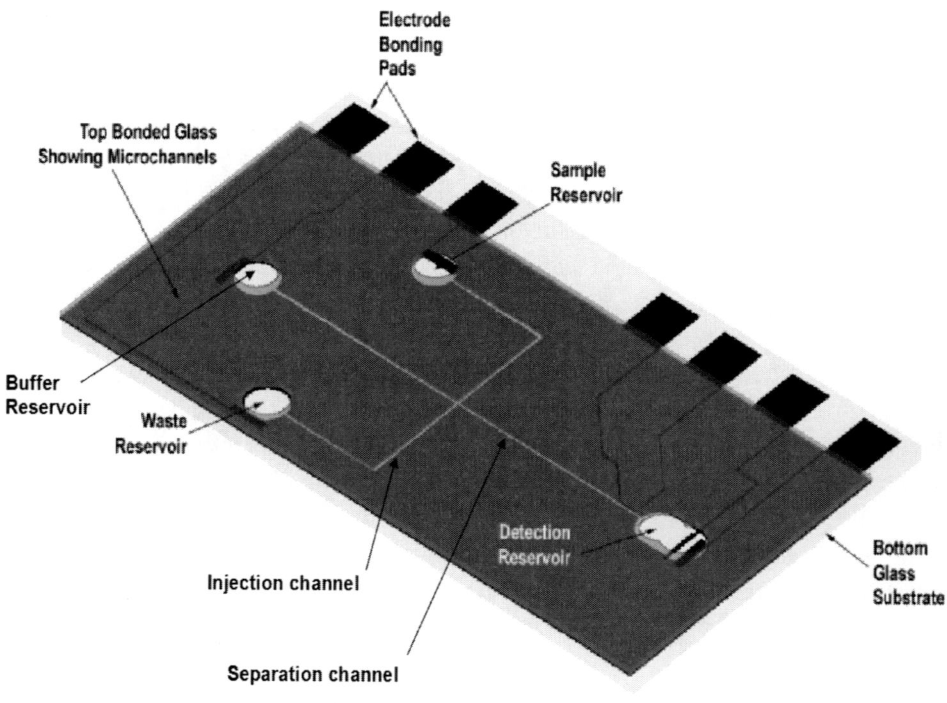

Figure 2. The layout of the capillary electrophoresis device with electrochemical detection. [Reprinted with permission from [29], copyright American Chemical Society.]

3. AC ELECTROPHORESIS

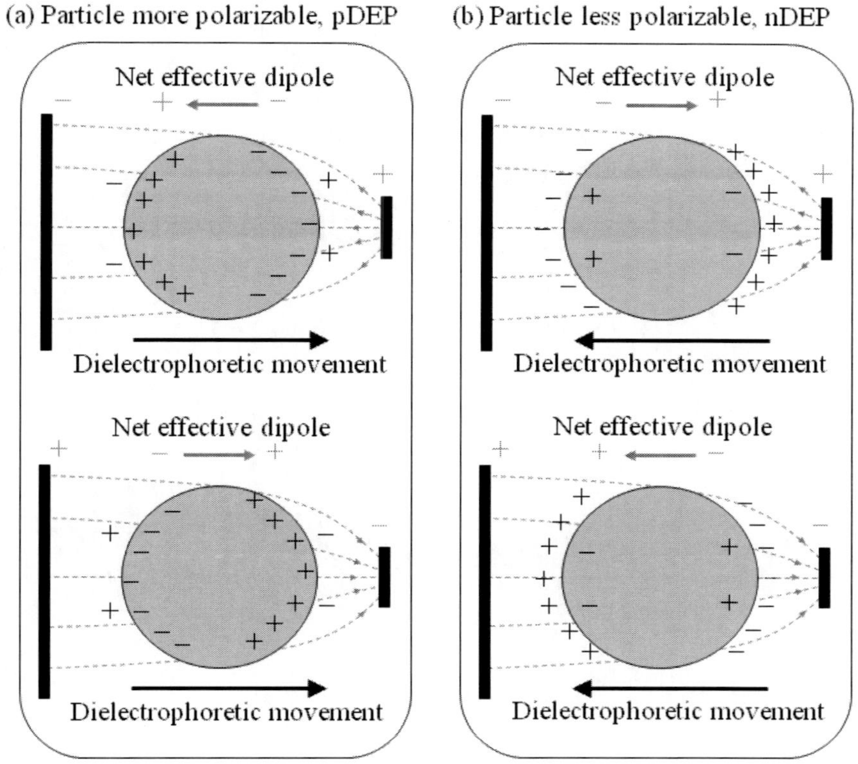

Figure 3. Schematic illustration of the dielectrophoretic movement for the two cases: (a) particle more polarizable then the medium, and (b) particle less polarizable then the medium. When the dipole is parallel to the electric field, the forces on the two poles are towards the nearest electrode. The pole facing the higher field strength region is pulled in that direction. Conversely, if the dipole is anti-parallel to the electric field, the forces on the two poles are repulsive. The pole facing the higher field strength has stronger repulsive force and repelled from that electrode.

AC electrophoresis, normally called dielectrophoresis (DEP), is a technique to manipulate dielectric biological particles suspended in fluid medium, such as bacteria and cells [32-35]. The particles are manipulated under a non-uniform AC electric field. The interaction between the induced dipole moment on the particle and the applied electric field generates force for manipulation [36, 37]. If the electric field is uniform, two poles of the dipole experience equal opposite forces and the particle has no movement. If the electric field is non-uniform, two forces are not equal and the particle moves. The direction of the movement depends on the polarizability of the particle, as illustrated in Figure 3. If the particle is more polarizable than the medium, more charge is induced on the inside of the interface. Both poles experience a force attracting them towards their facing electrode but the pole facing the higher field strength region has a stronger force. Therefore, the particle is pulled in the direction of increasing field strength. This is called positive dielectrophoresis (pDEP). Conversely, if the particle is less polarizable then the medium, more charge is induced on the outside of the interface. Both poles experience a repulsive force from the facing electrode but the pole facing the higher field strength region has a stronger force.

Therefore, the particle is pushed in the direction of decreasing field strength and negative dielectrophoresis (nDEP) occurs. Here, Clausius-Mossotti factor $\widetilde{f}_{CM}$ was introduced to describe a relaxation in the effective polarizability of the particle [36, 37]. The direction of the DEP force depends through the Clausius-Mossotti factor, on the permittivity and conductivity of both the particle and the suspending medium, and therefore on the frequency of the applied electric field. Mathematical description of DEP phenomenon has been investigated and interested readers can be referred to ref. [37].

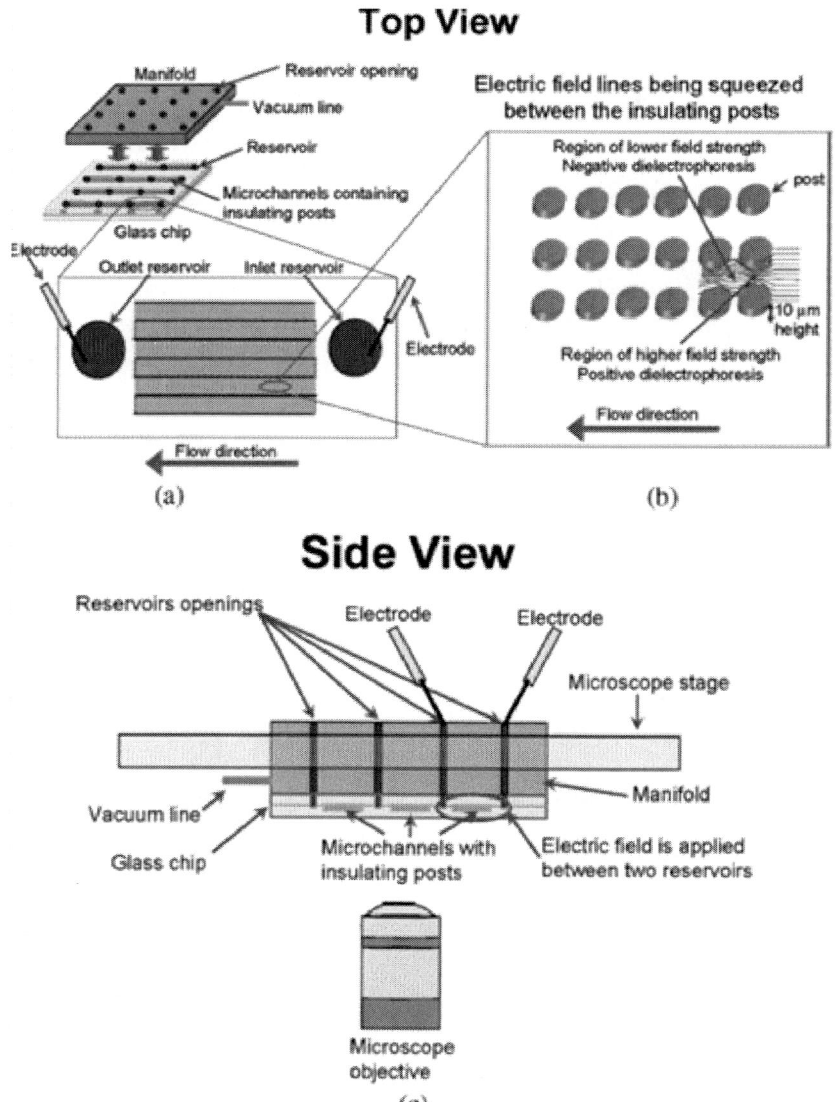

Figure 4. Experimental setup of the insulator-based DEP system. (a) Top view of the system showing the manifold, glass chip, and enlargement of the flow microchannel. (b) Illustration of the electric field lines being squeezed between the insulating posts. (c) Side view of the system mounted on the microscope stage. [Reprinted with permission from [33], copyright American Chemical Society.]

DEP have been reported to manipulate bacteria and cells in microfluidic systems. An example of concentrating and selectively releasing of live and dead *Escherichia coli* was demonstrated by applying stepped DC voltages across a microchannel containing an array of insulating posts [33]. This insulator-based DEP system avoided problems associated with the use of electrodes and illustration of the system is shown in Figure 4. By applying an electric voltage across the microchannel containing insulating posts, non-uniform electric field was generated among the posts for DEP manipulation. Cell trapping and release were controlled by modifying the relative DEP responses adjusted by the magnitude and frequency of the applied voltage. Another example of continuous DEP cell separation in a microfluidic chip was reported [34]. Three planar electrodes were fabricated in a separation channel for generation of the DEP forces. Therefore, positive DEP cells were moved away from the central streamline while the negative DEP cells remain in the central streamline (low field strength). Live and dead yeast cells were separated in the purity ranges of 95.9-97.3 and 64.5-74.3%, respectively. Moreover, using DEP technique as a selective filter was reported and removal of PCR inhibitors was demonstrated in microfluidic system [38]. To increase the sensitivity of the PCR, removal of PCR inhibitors in sample is essential. By using DEP, cells were captured by the electrodes in the fluidic chamber, while PCR inhibitors were washed by the buffer. After that, the DEP force can be turned off and the cells were released as a ready sample for the PCR.

4. DC ELECTROOSMOSIS

DC electroosmosis describes the fluidic movement across a charged surface due to an imposed DC electric field. Consider a solid surface is contacted with electrolyte medium. The electrostatic potential created by the net charge on the surface attracts ions of opposite charges (counterions) from the medium and repels ions with the same charges (co-ions). The density of the counterions is much higher then the density of co-ions in the region close to the surface and decreases away from the surface. This accumulation of the counterions close to the surface is called electrical double layer (EDL) [23]. If an electric field is applied tangential to the surface, the charges in the EDL experience a force and move. Therefore, the bulk fluid is pulled along due to the viscosity of the fluid. A flow is then generated with the velocity of zero immediately at the surface but maximum at the slip plane. This flow is called electroosmotic flow (EOF). Schematic illustration of the EOF in a cross-sectional microchannel is shown in Figure 5. Electroosmosis is therefore interpreted as a dynamic pump to manipulate fluids in low Reynold's number environment. Due to the principle of EOF, channel in high surface-to-volume ratio is required for efficient electroosmotic pumping. Mathematical understanding of the fluid dynamics of EOF is referred to ref. [23].

Numerous review papers summarized recent developments on the electroosmosis for pumping [39-41]. Demonstrations of the EOF in capillary channel are discussed below. For example, electroosmotic micropump was fabricated using 500-700 μm diameter fused-silica capillaries packed with 3.5 μm diameter non-porous silica particles [42]. The micropump achieved pressures in excess of 20 atm and flow rates of 3.6 μL/min for 2 kV applied potentials. There was no moving part in the electroosmotic micropump and the micropump could operate as both open (high flow rate) and closed (high pressure) systems. Also, a multistage electroosmotic micropump was developed for achieving higher output pressure [43]. The 1-3 stages electroosmotic micropump was fabricated using 100 mm × 320 μm internal diameter columns packed with 2 μm porous silica particles, fused-silica capillaries, and stainless electrodes. Compared with the 1-stage micropump, the output pressure of the 2- and 3-stage micropumps were two to three times higher, respectively. The flow rates of the 2- and 3-stage micropumps were identical with that of the 1-stage micropump at the same driving voltage. Electroosmotic micropumps have the advantages of no moving parts involved, quite operation, and bi-directional pumping controlled by switching the direction of the external electric field. The major limitation is the requirement of high voltage and electrically conductive solution.

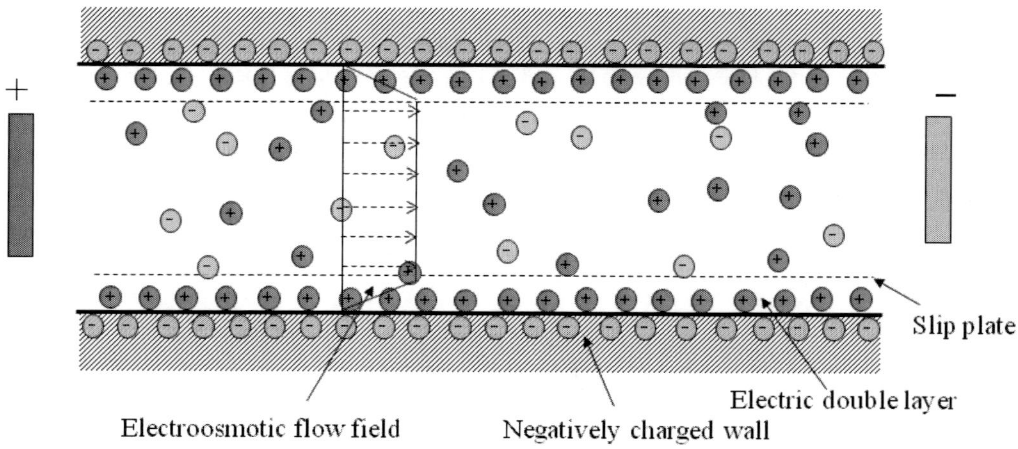

Figure 5. Schematic illustration of the electroosmotic flow in a cross-sectional microchannel.

Because of the mature developments of electroosmotic micropumps, various biological processes manipulated by EOF have been demonstrated recently [44, 45]. An electrokinetically controlled DNA hybridization microfluidic chip has been demonstrated and all processes from sample dispensing to hybridization detection can be completed in 5 min [44]. The chip was composed of a PDMS fluidic channel and a glass substrate with hybridization array, as shown in Figure 6. An H-type channel structure containing immobilized single-stranded oligonucleotide probes was utilized. The electroosmotic pumping can directly manipulate the controlled samples of nanoliter volume to the hybridization array and remove nonspecific adsorption. Hybridization, washing, and scanning procedures can be conducted simultaneously. Detection levels as low as 50 pM were recorded using an epifluorescence microscope. Alternatively, EOF can be utilized for the cell positioning to analyze cell-cell interactions [45]. The EOF chip was equipped with four electrodes to induce multi-directional EOF. Two cells were positioned in pocket-like microstructures by controlling the EOF direction. Cells were observed to have dye transfer through newly formed GJ channels between cells for up to 12 h after contact. This study took the advantage of the EOF to manipulate cells in a specific position. EOF provides constant and controllable hydrodynamic drag force near the substrate surface, while pressure-driven flow has a parabolic profile whose velocity is almost 0 near the substrate surface.

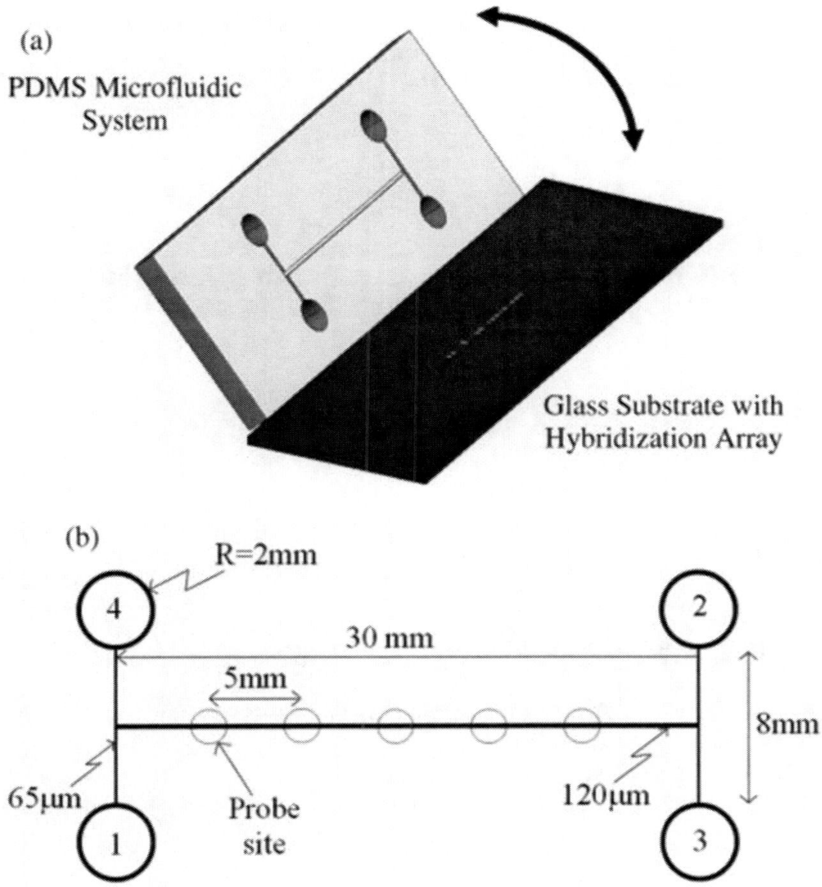

Figure 6. Illustration of the electrokinetically controlled DNA hybridization microfluidic chip. (a) Assembly procedure for PDMS fluidic channel and glass substrate with hybridization array. (b) H-type channel structure for DNA hybridization array chip: (1) sample port, (2) auxiliary port, (3) buffer port, and (4) wash port. [Reprinted with permission from [44], copyright American Chemical Society.]

5. AC Electroosmosis

Fluid flow generated by AC electroosmosis is induced by AC electric field through coplanar microelectrodes bathed in electrolyte medium. When an electric potential applies on the electrodes, charges at the electrode/electrolyte interface will accumulate on the electrode surface and form an EDL. Then, charges in the EDL interact with the tangential component of the electric field and experience a force. Hence, fluid is pulled along with the charges due to the viscosity of the fluid. The flow direction is constant because when the electric field direction changes, the sign of the induced charge also changes and therefore the direction of the force remains unchanged. This can generate a fluid velocity on the electrode surface, called slip velocity. This velocity depends on a number of factors: frequency, voltage, fluid conductivity, permittivity, viscosity, and importantly the scale of the system. Therefore, bulk fluid flow driven by slip velocity is generated. The illustration of bulk fluid flow generated by AC electroosmosis is shown in Figure 7.

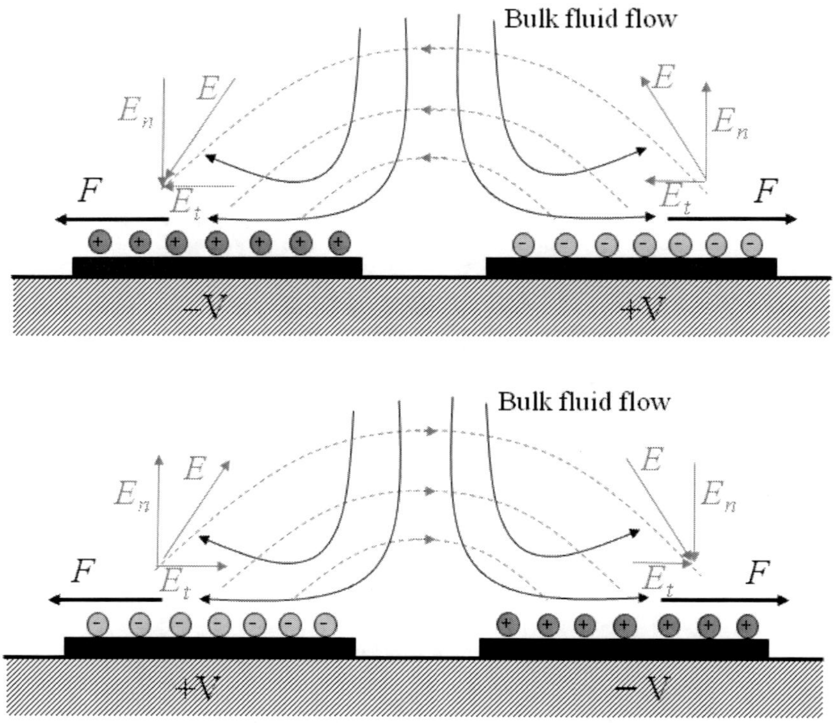

Figure 7. Bulk fluid flow generated by AC electroosmosis. Horizontal forces are induced by the interaction between the charges in the electric double layer and the tangential component of the applied AC electric field.

AC electroosmosis is a promising method of manipulating fluid with low operating voltage, which overcomes some of the issues of DC electroosmotic pumping. Asymmetric coplanar electrodes of different widths and gaps were demonstrated to pump fluids using AC electroosmosis [46]. Because of the asymmetry in the system, the tangential component of the electric field produces uneven force on the induced charges and rises to two different fluid velocities above the two different sized electrodes. Therefore, a unidirectional net flow can be generated in the microchannel. The AC electroosmotic pumping have been characterized by a number of groups. Theoretical analysis of the pumping phenomena was reported based upon an electroosmotic model in AC fields [47]. Also, an array of asymmetric indium tin oxide (ITO) interdigitated electrodes powered by AC voltages of less than 5 Vrms has been demonstrated to produce fluid flow in a microchannel [48]. Velocities of excess of 450 μm/s at 2.2 Vrms were achieved experimentally. It showed that the fluid velocity increases with the decreasing dimensions of the electrodes. Based on the AC electroosmosis phenomena, mixing enhancement utilizing coplanar meandering electrodes in a microchannel has been presented [49]. Various electrode configurations have been utilized to generate in-plane microvortices of clockwise, counter-clockwise, or pair rotation in a microchannel for mixing purpose. The mixing can be operated under the application of an AC signal with a voltage and frequency below 20 Vp-p and 10 kHz, respectively.

Interesting developments of using AC electroosmosis to manipulate biomolecules in microfluidic systems were demonstrated for specific biological application. Enhancement of heterogeneous immunoassays using AC electroosmosis was proposed to improve the rate of antibody binding to the sensor surface [50]. AC electroosmosis can generate rotational fluid flow near the sensor surface, i.e., electrodes, where the AC electric field was applied. Immunoassay was performed and showed the binding of secondary antibody on the sensor surface was significantly enhanced due to the possibility of local mixing. Enhancement of about 1.9 times at the center of the sensor and 6.7 times at the edges of the sensor was reported compared to a non-mixed counterpart. Moreover, concentration of single-stranded DNA (ssDNA) molecules in 20 nucleotides on a specific location has been developed [51]. A circular electrode pattern was utilized to concentrate ssDNA molecules in the bulk fluid. The combination of AC electroosmosis induced by AC electric field and DC electrophoresis induced by DC bias was applied simultaneously to hold the ssDNA molecules near the electrode surface. The applied potential was below 3 Vp-p and the ssDNA molecules were not shown damage.

6. Electrowetting

In the above discussion, fluid is handled by flowing continuously through microchannels. Alternatively, fluid can be also handled as droplet by droplet either in a closed volume or on an open surface. Droplet-based microfluidics has several advantages over traditional counterparts, such as elimination of dead volume, enhancement of mixing ratio, precise control of the volume, and encapsulation of biomolecules for monitoring. Individual droplets in a range between picoliters and microliters can be manipulated by the mechanism of electrowetting. Consider a droplet is placed on an electrode that is covered by a dielectric layer and then a hydrophobic layer, as shown in Figure 8. The contact angle of the droplet θ_0 on the hydrophobic layer is typically large. If a probe needle contacts the top of the droplet and an electric potential is applied between the probe and the electrode, the contact angle θ decreases depending on the voltage. The droplet is shown to wet the surface. The determination of the contact angle is based on the result of the mechanical equilibrium of a droplet resting on a solid surface [52]. This is the action of three surface tensions: γ_{LG} at the interface of the liquid and gas, γ_{SL} at the interface of the solid and liquid, and γ_{SG} at the interface of the solid and gas.

Therefore, the contact angle depends on the dielectric material, surface quality, liquid, temperature, and pressure. When an electric potential is applied across the solid-liquid interface, EDL is formed in the interface causing the surface tension γ_{SL} decreases with a rate of charge density.

The contact angle decreases in order to have mechanical equilibrium of three surface tensions. The droplet spreads on the solid surface, as illustrated in Figure 8. This phenomenon is generally called electrowetting. In this case, the droplet is placed on an electrode that is covered by a dielectric layer and then a hydrophobic layer.

A thin-film dielectric layer can sustain much higher voltage for generating more contact angle changes. The term electrowetting-on-dielectric (EWOD) is used to clarify that the electrowetting is induced on a dielectric material rather than metal.

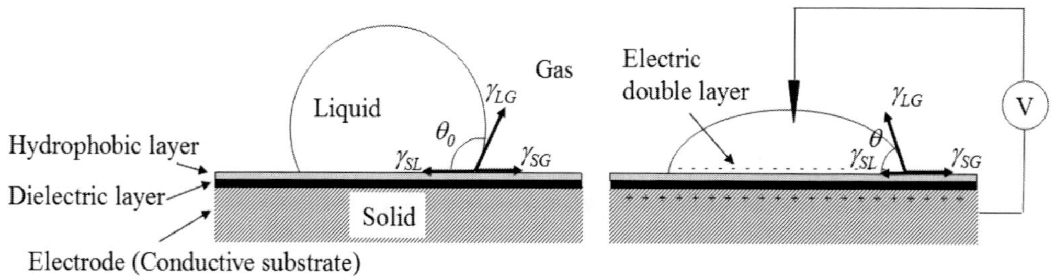

Figure 8. Demonstration of the EWOD principle with a droplet. (Left) The droplet is placed on the hydrophobic layer. The contact angle θ_0 is typically large. (Right) When an electric potential is applied between the probe needle and the electrode, the contact angle θ decreases and wet the surface.

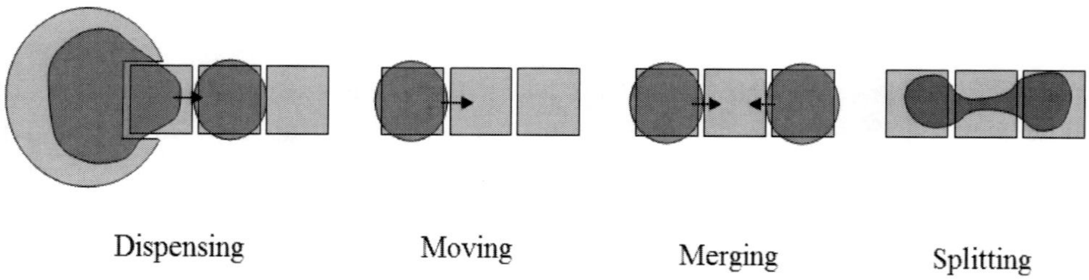

Figure 9. Illustration of the basic microfluidic functions performed by EWOD.

Based on the electrowetting phenomenon, manipulation of droplets can be achieved by an array of driving electrodes. Basic microfluidic functions, such as dispensing, moving, merging, and splitting, were demonstrated [53-56]. Illustrations are shown in Figure 9. By applying electrical potentials to sequential electrodes, a droplet can be dispensed from a reservoir, transported to any position on the array, merged with other droplets to perform reactions, and split into two droplets. This droplet-based system was suggested that the configuration is easier than the microchannel-based system. The flexibility of these electrowetting-based systems makes them suitable for a wide variety of applications. For example, a microfluidic device based on electrowetting manipulation has been developed to perform multiplexed enzyme analyses [57]. Samples and reagents in the form of discrete droplets were manipulated on the device. Droplets of alkaline phosphatase and fluorescein diphosphate were merged and mixed on the device, and then the fluorescent product was detected by fluorescence plate reader. The detection limit was achieved ~7.0×10^{-20} M. Also, heterogeneous immunoassays have been demonstrated by efficient handling of magnetic microbeads using electrowetting manipulation [58]. A sample droplet and a reagent droplet containing magnetic beads conjugated to primary capture antibodies, blocking proteins, and secondary antibodies were dispensed on the system. These two droplets were then merged, mixed, and incubated by electrowetting manipulation. A permanent magnet was applied to immobilize the sandwiched microbead complexes, followed by the washing of the unbound components. Finally, a reagent droplet was applied for the chemiluminescent detection.

Sandwich heterogeneous immunoassays on human insulin and interleukin-6 (IL-6) were demonstrated with a total time to result of 7 min for each assay. Moreover, the electrowetting manipulation has been applied to the PCR for potential point-of-care applications [59]. PCR in the droplets showed amplification efficiencies with no evaporation loss. The optimal hold time was found to be 9 s and 30 s for denaturation and annealing/extension in thermal cycling, respectively. Droplet-based PCR can be monitored in real time and provides amplification with a cycle threshold of ~10 cycles earlier than the benchtop instruments.

7. OPTO-ELECTROKINETICS

As discussed above, electrokinetic forces can be induced by microelectrodes fabricated on a substrate in microfluidic systems. Recently, electrokinetic forces have been also demonstrated to be induced by opto-electric phenomenon. Critical reviews of opto-electrokinetic manipulation in microfluidic systems were reported [60, 61]. Schematic illustrations of the opto-electrokinetic manipulation are shown in Figure 10. Typically, there are two approaches for opto-electrokinetic manipulation: (1) direct change of fluid properties by light and (2) change of surface conductivity by light. In typical experimental setup, working fluid is placed in the gap between two parallel electrodes for the operations. In the case of former approach, local temperature is increased by illuminating a strong light source. Hence, change of electrical conductivity and permittivity of the fluid is induced to generate electrohydrodynamic vortices under an electric field applied across the electrodes. By using this approach, opto-electrokinetic manipulation and concentration of colloidal particles was demonstrated [62, 63]. Particle manipulation was performed between two parallel ITO electrodes biased with an AC signal and illuminated with near-infrared (1064nm) laser. Highly localized laser induced a highly non-uniform heating of the electrode surface. Electrohydrodynamic vortices were generated to transport particles toward the illuminated site. Polystyrene particles with diameters ranging from 2.0 to 0.1 μm suspended in a low conductivity solution (2.0 mS/m) could be aggregated at specific locations. In the case of latter approach, a photoconductive layer is coated on one of the electrodes. Opto-electrokinetic forces are generated by the change of the surface conductivity of the photoconductive layer induced by light illumination. Partially illuminated area of the surface becomes more conductive than other area. Hence, non-uniform electric field in the fluid is formed and results electrokinetic phenomena. Therefore, a weak conventional white light source is applicable for inducing electric fields, such as digital micromirror device (DMD) [64], liquid crystal display (LCD) [65], and beam projector [66]. The use of such kind of light sources provides a higher degree of freedom to construct electric fields through illumination. Massively parallel manipulation of single cells and microparticles was demonstrated using a light-emitting diode operating at a wavelength of 625 nm [64]. The optical images shown on the DMD were focused onto the photoconductive surface and created the non-uniform electric field for DEP manipulation. The minimum optical intensity required to induce the electric field was 10 nW/μm, which is 100,000 times lower than that used in optical tweezers. Using incoherent optical images for manipulation can control the DEP forces over a large area. Alternatively, microparticles were manipulated by opto-electrokinetic forces on a LCD [65]. The system was constructed on a LCD and the LCD image directly formed virtual electrodes on the photoconductive layer without lens or optical alignments. The LCD module with light source brighter than 5,000 lux was utilized for the manipulation of 45 μm polystyrene beads. Moreover, cell manipulation has been demonstrated using an optically addressable projection array of 800 by 600 [66]. A 1,200 lm commercial TFT-LCD projector was used to produce light source on the system. Different virtual electrode patterns can be programmed by a computer. Parallel manipulation of cells without interfering each other was shown. The technique of opto-electrokinetic manipulation opens new portals of dynamic, multi-scale, and high-throughput manipulation of biomolecules.

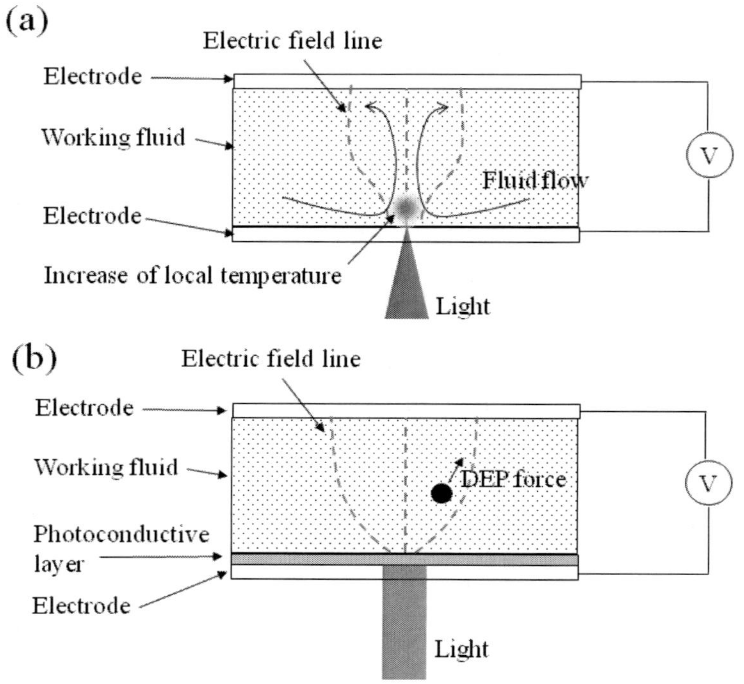

Figure 10. Schematic illustration of the opto-electrokinetic particle manipulation. (a) Direct change of fluid properties by light. (b) Change of surface conductivity by light.

CONCLUSION

Electrokinetics provides a simple approach for the manipulation of biomolecules. By designing different electrode patterns, different kinds of biomolecules can be manipulated by different electrokinetic forces. This chapter reviewed the most commonly used electrokinetic phenomena, i.e., DC electrophoresis, AC electrophoresis, DC electroosmosis, AC electroosmosis, electrowetting, and opto-electrically induced electrokinetics. Electrokinetic manipulation is efficient because the length scale of manipulation matches various biomolecules. Since the development of biomedical applications using electrokinetic manipulation is still ongoing, we expect more and more practical demonstrations will be presented in the near future.

REFERENCES

[1] D.J. Harrison, K. Fluri, K. Seiler, Z. Fan, C.S. Effenhauser, A. Manz, "Micromachining a miniaturized capillary electrophoresis-based chemical analysis system on a chip", *Science* 261:895-8973, 1993.

[2] T. Thorsen, S.J. Maerkl, S.R. Quake, "Microfluidic large-scale integration", *Science* 298:580-584, 2002.

[3] S. Park, Y. Zhang, S. Lin, T.H. Wang, S. Yang, "Advances in microfluidic PCR for point-of-care infectious disease diagnostics", *Biotechnology Advances* 29:830-839, 2011.

[4] A.H. Diercks, A. Ozinsky, C.L. Hansen, J.M. Spotts, D.J. Rodriguez, A. Aderem, "A microfluidic device for multiplexed protein detection in nano-liter volumes", *Analytical Biochemistry* 386:30-35, 2009.

[5] F. Amirouche, Y. Zhou, T. Johnson, "Current micropump technologies and their biomedical applications", *Microsystem Technologies* 15:647-666, 2009.

[6] N.C. Tsai, C.Y. Sue, "Review of MEMS-based drug delivery and dosing systems", Sensors and Actuators A: *Physical* 134:555-564, 2007.

[7] H.T.G. Van Lintel, F.C.M. van de Pol, S. Bouwstra, "A piezoelectric micropump based on micromachining of silicon", *Sensors and Actuators* 15:153-167, 1988.

[8] J.G. Smits, "Piezoelectric micropump with three valves working peristaltically", Sensors and Actuators A: *Physical* 21:203-206, 1990.

[9] J. Kan, Z. Yang, T. Peng, G. Cheng, B. Wu, "Design and test of a high-performance piezoelectric micropump for drug delivery", Sensors and Actuators A: *Physical* 121:156-161, 2005.

[10] I. Izzo, D. Accoto, A. Menciassi, L. Schmitt, P. Dario, "Modeling and experimental validation of a piezoelectric micropump with novel no-moving-part valves", Sensors and Actuators A: *Physical* 133:128-140, 2007.

[11] M.A. Unger, H.P. Chou, T. Thorsen, A. Scherer, S.R. Quake, "Monolithic microfabricated valves and pumps by multilayer soft lithography", *Science* 288:113-116, 2000.

[12] Y.N. Yang, S.K. Hsiung, G.B. Lee, "A pneumatic micropump incorporated with a normally closed valve capable of generating a high pumping rate and high back pressure", *Microfluidics and Nanofluidics* 6:823-833, 2009.

[13] W. Inman, K. Domansky, J. Serdy, B. Owens, D. Trumper, L.G. Griffith, "Design, modeling and fabrication of a constant flow pneumatic micropump", *Journal of Micromechanics and Microengineering* 17:891-899, 2007.

[14] S.M. Ha, W. Cho, Y. Ahn, "Disposable thermo-pneumatic micropump for bio lab-on-a-chip application", *Microelectronic Engineering* 86:1337-1339, 2009.

[15] O.C. Jeong, S. Konishi, "Fabrication and drive test of pneumatic PDMS micro pump", Sensors and Actuators A: *Physical* 135:849-856, 2007.

[16] R. Zengerle, J. Ulrich, S. Kluge, M. Richter, A. Richter, "A bidirectional silicon micropump", Sensors and Actuators A: *Physical* 50:81–86, 1995.

[17] M.M. Teymoori, E. Abbaspour-Sani, "Design and simulation of a novel electrostatic peristaltic micromachined pump for drug delivery applications", Sensors and Actuators A: *Physical* 117:222-229, 2005.

[18] O. Francais, I. Dufour, "Dynamic simulation of an electrostatic micropump with pull-in and hysteresis phenomena", Sensors and Actuators A: *Physical* 70:56-60, 1998.

[19] C. Yamahata, C. Lotto, E. Al Assaf, M.A.M. Gijs, "A PMMA valveless micropump using electromagnetic actuation", Microfluidics and Nanofluidics 1:197–207, 2005.

[20] K. Weber, M. Osborn, "The reliability of molecular weight determinations by dodecyl sulfate-polyacrylamide gel electrophoresis", *Journal of Biological Chemistry* 244:4406-4412, 1969.

[21] D.M. Neville, "Molecular weight determination of protein-dodecyl sulfate complexes by gel electrophoresis in a discontinuous buffer system", *Journal of Biological Chemistry* 246:6328-6334, 1971.

[22] M. Orita, H. Iwahana, H. Kanazawa, K. Hayashi, T. Sekiya, "Detection of polymorphisms of human DNA by gel electrophoresis as single-strand conformation polymorphisms", *Proceedings of the National Academy of Sciences of the United States of America* 86:2766-2770, 1989.

[23] D. Li, *Electrokinetics in Microfluidics*, Academic Press, Oxford, 2004.

[24] W.R. Vandaveer IV, S.A. Pasas, R.S. Martin, S.M. Lunte, "Recent developments in amperometric detection for microchip capillary electrophoresis", *Electrophoresis* 23:3667-3677, 2002.

[25] S. Mouradian, "Lab-on-a-chip: applications in proteomics", Current *Opinion in Chemical Biology* 6:51-56, 2002.

[26] N.A. Lacher, K.E. Garrison, R.S. Martin, S.M. Lunte, "Microchip capillary electrophoresis/ electrochemistry", *Electrophoresis* 22:2526-2536, 2001.

[27] K. Swinney, D.J. Bornhop, "Detection in capillary electrophoresis", *Electrophoresis* 21:1239-1250, 2000.

[28] J. Wang, "Electrochemical detection for capillary electrophoresis microchips: A review", *Electroanalysis* 17:1133-1140, 2005.

[29] R.P. Baldwin, T.J. Roussel, M.M. Crain, V. Bathlagunda, D.J. Jackson, J. Gullapalli, J.A. Conklin, R. Pai, J.F. Naber, K.M. Walsh, R.S. Keynton, "Fully integrated on-chip electrochemical detection for capillary electrophoresis in a microfabricated device", *Analytical Chemistry* 74:3690-3697, 2002.

[30] A.T. Woolley, D. Hadley, P. Landre, A.J. deMello, R.A. Mathies, M.A. Northrup, "Functional integration of PCR amplification and capillary electrophoresis in a microfabricated DNA analysis device", *Analytical Chemistry* 68:4081-4086, 1996.

[31] M. Yu, Q. Wang, J.E. Patterson, A.T. Woolley, "Multilayer polymer microchip capillary array electrophoresis devices with integrated *on-chip labeling for high-throughput protein analysis",* Analytical Chemistry 83:3541-3547, 2011.

[32] G.H. Markx, P.A. Dyda, R. Pethig, "Dielectrophoretic separation of bacteria using a conductivity gradient", *Journal of Biotechnology* 51:175-180, 1996.

[33] B.H. Lapizco-Encinas, B.A. Simmons, E.B. Cummings, Y. Fintschenko, "Dielectrophoretic concentration and separation of live and dead bacteria in an array of insulators", *Analytical Chemistry* 76:1571-1579, 2004.

[34] I. Doh, Y.H. Cho, "A continuous cell separation chip using hydrodynamic dielectrophoresis (DEP) process", *Sensors and Actuators A: Physical* 121:59-65, 2005.

[35] D.S. Gray, J.L. Tan, J. Voldman, C.S. Chen, "Dielectrophoretic registration of living cells to a microelectrode array", *Biosensors and Bioelectronics* 19:771-780, 2004.

[36] H.A. Pohl, *Dielectrophoresis*, Cambridge University Press, Cambridge, 1978.

[37] A. Ramos, *Electrokinetics and Electrohydrodynamics in Microsystems*, Springer, New York, 2011.
[38] I.R. Perch-Nielsen, D.D. Bang, C.R. Poulsen, J. El-Ali, A. Wolff, "Removal of PCR inhibitors using dielectrophoresis as a selective filter in a microsystem", *Lab on a Chip* 3:212-216, 2003.
[39] B.D. Iverson, S.V. Garimella, "Recent advances in microscale pumping technologies: a review and evaluation", *Microfluidics and Nanofluidics* 5:145-174, 2008.
[40] C.C. Chang, R.J. Yang, "Electrokinetic mixing in microfluidic systems", *Microfluidics and Nanofluidics* 3:501-525, 2007.
[41] R. Chein, J. Liao, "Analysis of electrokinetic pumping efficiency in capillary tubes", *Electrophoresis* 28:635-643, 2007.
[42] S. Zeng, C.H. Chen, J.C. Mikkelsen, J.G. Santiago, "Fabrication and characterization of electroosmotic micropumps", *Sensors and Actuators B: Chemical* 79:107–114, 2001.
[43] L. Chen, H. Wang, J. Ma, C. Wang, Y. Guan, "Fabrication and characterization of a multi-stage electroosmotic pump for liquid delivery", Sensors and Actuators B: *Chemical* 104:117–123, 2005.
[44] D. Erickson, X. Liu, U. Krull, D. Li, "Electrokinetically controlled DNA hybridization microfluidic chip enabling rapid target analysis", *Analytical Chemistry* 76:7269-7277, 2004.
[45] K. Terao, Y. Kitazawa, R. Yokokawa, A. Okonogi, H. Kotera, "Open-access and multi-directional electroosmotic flow chip for positioning heterotypic cells", *Lab on a Chip* 11:1507-1512, 2011.
[46] A.B.D. Brown, C.G. Smith, A.R. Rennie, "Pumping of water with ac electric fields applied to asymmetric pairs of microelectrodes", *Physical Review E* 63:016305, 2000.
[47] A. Ramos, A. Gonzalez, A. Castellanos, N.G. Green, H. Morgan, "Pumping of liquids with ac voltages applied to asymmetric pairs of microelectrodes", *Physical Review E* 67:056302, 2003.
[48] M. Mpholo, C.G. Smith, A.B.D. Brown, "Low voltage plug flow pumping using anisotropic electrode arrays", *Sensors and Actuators B* 92 262-268, 2003.
[49] S.H. Huang, S.K. Wang, H.S. Khoo, F.G. Tseng, "AC electroosmotic generated in-plane microvortices for stationary or continuous fluid mixing", Sensors and Actuators B: *Chemical* 125:326-336, 2007.
[50] R. Hart, R. Lec, H. Noh, "Enhancement of heterogeneous immunoassays using AC electroosmosis", Sensors and Actuators B: *Chemical* 147:366-375, 2010.
[51] K.F Lei, H. Cheng, K.Y. Choy, L.M.C. Chow, "Electrokinetic DNA concentration in microsystems", Sensors and Actuators A: *Physical* 156:381-387, 2009.
[52] A.W. Adamson, *Physical Chemistry of Surface*, Wiley, New York, 1990.
[53] M.W.L. Watson, M. Abdelgawad, G. Ye, N. Yonson, J. Trottler, A.R. Wheeler, "Microcontact printing-based fabrication of digital microfluidic devices", *Analytical Chemistry* 78:7877-7885, 2006.
[54] O.D. Velev, B.G. Prevo, K.H. Bhatt, "On-chip manipulation of free droplets", *Nature* 426:515-516, 2003.
[55] MG. Pollack, A.D. Shenderov, R.B. Fair, "Electrowetting-based actuation of droplets for integrated microfluidics", *Lab on a Chip* 2:96-101, 2002.

[56] J. Lee, H. Moon, J. Fowler, T. Schoellhammer, C.J. Kim, "Electrowetting and electrowetting-on-dielectric for microscale liquid handling", Sensors and Actuators A: *Physical* 95:259-268, 2002.

[57] E.M. Miller, A.R. Wheeler, "A digital microfluidic approach to homogeneous enzyme assays", Analytical Chemistry 80:1614-1619, 2008.

[58] R.S. Sista, A.E. Eckhardt, V. Srinivasan, M.G. Pollack, S. Palanki, V.K. Pamula, "Heterogeneous immunoassays using magnetic beads on a digital microfluidic platform", Lab *on a Chip* 8:2188-2196, 2008.

[59] F. Wang, M.A. Burns, "Performance of nanoliter-sized droplet-based microfluidic PCR", *Biomedical Microdevices* 11:1071-1080, 2009.

[60] A. Kumar, S.J. Williams, H.S. Chuang, N.G. Green, S.T. Wereley, "Hybrid opto-electric manipulation in microfluidics - opportunities and challenges", *Lab on a Chip* 11:2135-2148, 2011.

[61] H. Hwang, J.K. Park, "Optoelectrofluidic manipulation of nanoparticles and biomolecules", *Advances in Optoelectronics*, 2011:482483, 2011.

[62] S.J. Williams, A. Kumar, S.T. Wereley, "Electrokinetic patterning of colloidal particles with optical landscapes", *Lab on a Chip* 8:1879-1882, 2008.

[63] A. Kumar, J.S. Kwon, S.J. Williams, N.G. Green, N.K. Yip, S.T. Wereley, "Optically modulated electrokinetic manipulation and concentration of colloidal particles near an electrode surface", *Langmuir* 26(7):5262-5272, 2010.

[64] P.Y. Chiou, A.T. Ohta, M.C. Wu, "Massively parallel manipulation of single cells and microparticles using optical images", *Nature* 436:370-372, 2005.

[65] W. Choi, S.H. Kim, J. Jang, J.K. Park, "Lab-on-a-display: a new microparticle manipulation platform using a liquid crystal display (LCD)", *Microfluidics and Nanofluidics* 3:217-225, 2007.

[66] Y.S. Lu, Y.P. Huang, J.A. Yeh, C. Lee, Y.H. Chang, "Controllability of non-contact cell manipulation by image dielectrophoresis (iDEP)", *Optical and Quantum Electronics* 37:1385-1395, 2005.

INDEX

#

A

B

C

D

E

F

G

H

I

J

K

L

M

N

O

P

Q

R

S

T

U

V

W

X

Z